From Girls in Their Elements to Women in Science

"An original and incisively written account of memory, memory-work with nature, and the creative spaces in which girls discover their own power as they discover themselves and others. This is a unique and inspiring study of personal science grounded in experiences with and becoming a significant science rooted in lived life."

Maxine Greene, Professor of Philosophy
and Education (Emerita), Teachers College, Columbia University

"…a fascinating study of the use of memories about nature and scientific thought to analyze the process of our acculturation to masculinist, Western science and, even more importantly, the forms of our resistance to the forces of socialization that allow us to transcend societal constraints. Judith Kaufman and colleagues have produced a highly original book that illustrates the power of personal and political transformation through collective feminist process. I can see using this book in a research seminar for graduate and advanced undergraduate students. It provides a compelling model for opening spaces for transforming our knowledge and our lives."

Bonnie B. Spanier, Associate Professor of Women's Studies,
State University of New York at Albany

"…contributes to the growing body of literature about [some] girls' unwillingness to study or to "do" science the way it is currently learned about and practiced…. Also of value are the authors' insights as to an ideal dynamic in science education: from guided participation, through apprenticeship, through participatory appropriation, which I take to mean, feeling equal to and powerful enough to study the natural world on one's own."

Sheila Tobias, Author and Science Education Consultant

From Girls in Their Elements to Women in Science

Studies in the
Postmodern Theory of Education

Joe L. Kincheloe and Shirley R. Steinberg
General Editors

Vol. 116

PETER LANG
New York • Washington, D.C./Baltimore • Bern
Frankfurt am Main • Berlin • Brussels • Vienna • Oxford

Judith S. Kaufman, Margaret S. Ewing,
Diane M. Montgomery, Adrienne E. Hyle,
and Patricia A. Self

From Girls in Their Elements
to Women in Science

Rethinking Socialization
through Memory-Work

PETER LANG
New York • Washington, D.C./Baltimore • Bern
Frankfurt am Main • Berlin • Brussels • Vienna • Oxford

Library of Congress Cataloging-in-Publication Data

From girls in their elements to women in science:
rethinking socialization through memory-work / Judith S. Kaufman…[et al.].
p. cm. — (Counterpoints; vol. 116)
Includes bibliographical references and index.
1. Memory—Social aspects. 2. Human information processing—
Social aspects. 3. Cognition and culture. 4. Social perception.
5. Women scientists—Psychology. 6. Socialization.
I. Kaufman, Judith S. II. Counterpoints (New York, N.Y.); vol. 116.
BF378.S65 F76   153.1'2'082—dc21   99-053315
ISBN 0-8204-4512-6
ISSN 1058-1634

Die Deutsche Bibliothek-CIP-Einheitsaufnahme

From girls in their elements to women in science:
rethinking socialization through memory-work / Judith S. Kaufman…
–New York; Washington, D.C./Baltimore; Bern;
Frankfurt am Main; Berlin; Brussels; Vienna; Oxford: Lang.
(Counterpoints; Vol. 116)
ISBN 0-8204-4512-6

Cover design by Joni Holst

The paper in this book meets the guidelines for permanence and durability
of the Committee on Production Guidelines for Book Longevity
of the Council of Library Resources.

© 2003 Peter Lang Publishing, Inc., New York
275 Seventh Avenue, 28th Floor, New York, NY 10001
www.peterlangusa.com

Printed in the United States of America

Dedicated to Angie and her legacy in BreAnn

Contents

Acknowledgments   ix

Chapter 1   Introduction   1

PART ONE

Chapter 2   Nature to Natural Science   13

Chapter 3   Memory and Memory-Work   25

PART TWO

Chapter 4   Making Sense   45

Chapter 5   Metaphor: Girls in Their Elements   66

Chapter 6   Making New Meaning: Creative Acts   85

Chapter 7   Family Landscapes in Nature   99

Chapter 8   The Power of Girls   116

Chapter 9   Interruptions   132

Illustrations   141

Appendix A   147

Appendix B   149

References   153

Index   161

Acknowledgments

We appreciate the assistance of many people in the writing of this book. The following people read drafts and provided critiques, a number of them particularly detailed, pointed, and enormously useful: Rosebud Elijah, Sidney, Holly, Leah and Ann Ewing, Katherine Kocan, Renée McBride, Jan McDonald, Leslie Miller, and Celia Oyler. Discussions with D. Dutt, Larisa Self, and Angela Self Redcross sharpened our perception and broadened our understanding of several themes developed here. Jan McDonald posed critical questions that pushed us to consider ideas more deeply and describe them with greater clarity. Additional editorial assistance was ably provided by Colleen McLaughlin, Michelle Edwards, and Kerri Kearney.

We thank Roe Balian for her energy and persistence in transcribing audio-tapes of our research meetings. Funds for transcription were provided by the Oklahoma State University College of Education and the School of Education and Allied Human Services at Hofstra University.

Personal support over the long life of this project has recharged, countless times, our enthusiasm, patience, and stamina, to say nothing of our sense of humor. Angela, Grover, Jan, Larisa, Ross, and Sidney: We thank you.

The lines from "Transcendental Etude," from *The Fact of a Doorframe: Poems Selected and New, 1950–1984* by Adrienne Rich. Copyright © 1984 by Adrienne Rich. Copyright © 1975, 1978 by W. W. Norton & Company, Inc. Copyright © 1981 by Adrienne Rich. Used by permission of the author and W. W. Norton & Company, Inc.

Some excerpts originally appeared in "The Hard Work of Remembering: Memory Work as Narrative Research" by M. Ewing, A. Hyle, J. Kaufman, D. Montgomery and P. Self. In *Feminist Empirical Research* edited by Joanne Addison and Sharon James McGee. Published in 1999 by Boynton/Cook Publishers, Inc., a division of Reed Elsevier, Portsmouth, NH.

CHAPTER 1

Introduction

Bell, Adolescent, Fire~Sweaty Palms Memory: *She has more of an impression than an actual memory, but the first thought that was triggered was an image of the Bunsen burner on the lab table in her 9th grade honors biology class. That's it—but the burner serves as a magnet for all the negative emotions surrounding that class. She had heard that the teacher, Mr. Wilson, terrorized his students. She never knew if he would call on her to perform a piece of an activity or answer a question, and god help her if she didn't do it just right or have the correct answer for his question. He needled the students, embarrassed them, maliciously teased them—unless of course it was a student he liked, and then he displayed a sort of kindness that was belied by his unpredictability. You never knew when he would explode at one student or all of them.*

She remembers the slate lab tables and the stools and coming in to class once the bell rang. The image of the gradually drying wet marks made by her sweating palms as she pressed and repressed them on the table. There's also a sense that she was unsure of the Bunsen burners. She was not quite sure how they worked or how they were lit—she had to relearn how to work with one each time she had to light it—and each relearning made her vulnerable to a possible attack by Mr. Wilson. The next year she had heard that Mr. Wilson's college-age son had committed suicide by hanging himself in a closet at home.

We toyed with the idea of naming this book "Sweaty Palms." All of us had similar anxieties and experiences with science in school, and even now, as professionals who are successful in our academic environments, we still struggle with our roles in the social and natural sciences. The issues are different now; they have less to do with feelings of worthlessness and self-doubt and more to do with dissatisfaction with the traditional scholarship and research of our respective disciplines. At the time we began this project, we felt intellectually isolated to one degree or another. In search of an intellectual and emotional community of colleagues, Judy convened our group with the prospect of engaging in some radically different research. We were willing to take intellectual risks, cross disciplinary boundaries, and challenge our belief systems. We were also willing to eschew the immediate academic payoffs of publications, presentations, and grants. We began this project in 1994.

At Judy's urging we chose an intriguing and not widely known methodology called memory-work. Essentially, the method allows a group of people, engaged collectively, to examine how they have been socialized and how they participate in their socialization within a culture. The data generated by the method are the memories of the members of the group triggered by cues or words related to the particular area of socialization the group wants to study. The aim of memory-work is consciousness of how we shape and are shaped: how we see, feel, and think about the world in very particular ways.

The work of two feminist research collectives guided our efforts. The first is the research of a German group described by Haug (1987), which originally developed memory-work; the second is the work of an Australian group (Crawford, Kippax, Onyx, Gault, & Benton, 1992). Haug (1987) chronicles a "collective's attempts to analyze women's socialization by writing stories out of their own personal memories" (p. 13). Among the topics the collective studied were parts of the body such as legs or characteristics such as body hair. The members wrote stories of their earliest memories, and, through collective analysis of the memories, they found that their socialization could in part be described as a process by which the female body becomes sexualized. That collective used memory-work "as a bridge to span the gap between theory and their personal experience" (1987, p. 14). The members of the Australian group (Crawford et al., 1992) studied emotions in their earliest memories and, based on collective analysis, found that their social construction (and participation in that construction) as young girls was partly a function of how they learned to define emotion terms like anger and happiness.

At our first meeting, we discussed a variety of ideas as possibilities for memory-work investigation, and we chose to use memory-work to examine our socialization in relationship to nature and science. Our question, which changed over the course of this project, was "What can memory-work tell us about our relationship to nature and therefore to science?" We decided to use the classical elements (air, earth, fire, and water) as cues for our memories. We reasoned that the memories cued by the elements would help us examine our socialization in relation to the natural world. Insights into our relationship to nature would in turn, we hoped, give us insight into our relationship to science because science is tied to the observation of nature. Early on we expanded the cues to include tree; as an organism it is a cue more closely connected with the living world than are the classical elements. We began by recalling our earliest childhood memories and then decided to include adolescent as well as adult memories.

Following Crawford et al.'s (1992) strategies for using memory-work, we read the memories aloud, asked questions, clarified details present and missing, and added needed context. From a corpus of over 90 memories generated in response to the cues earth, air, water, fire, and tree, we ultimately drew conclusions about what we were taught to value in nature and what we forgot, often because it was not valued by society. Multiple themes and perspectives emerged in our analysis. These were related to power, family, metaphor, creativity, and the sensuous. These perspectives provided insight into the shape of our relationship to nature.

Through our analysis of the childhood, adolescent, and adult experiences in nature described in our memories, we also discovered what we have come to call our "personal science." This science is embodied in the sense that the experience is located personally in our bodies. It often has a strongly sensuous aspect and a lack of distance from the natural world. For children, experience of the natural world, in particular, produces "a sense of some profound continuity with natural processes" (Cobb, 1959, p. 538). The science that emerges from such experience is knowledge of nature acquired in the broad contexts of home, play, and human relationships, contexts in which subjectivity is integral. Sometimes we acquire this knowledge systematically over years, though it may be called child's play or common sense by observers. Nevertheless, if "science is the acquisition of reliable but not infallible knowledge of the real world" (Strahler, 1992, p. 8), a sandbox may provide a setting for science. For example, the following early memory of a sandbox shows knowledge of the layers of sand on a hot summer day, gained through personal science.

Sue, Child, Earth~Sandbox Memory: *The top of the sand is hot and dry, not uncomfortable, but definitely not wonderfully cool and wet like the sand that is down deeper in the sandbox. The difference between the two, the top and the bottom, is dissipated as the children mix the sand from the top and the sand from the bottom together. However, no matter how much bottom sand they mix with the top sand, it is never quite wet enough to "play with."*

As a result of this project, we have altered our relationship to traditional science. While we were all critical of a science enveloped by empiricism and positivism, memory-work enabled us to bring science to a personal level of experience. Through our memories we were able to experience how we shape and are shaped by a set of beliefs that prioritizes one way of knowing the world above others. We recognize that the stories we have reconstructed through memory-work are only a few among a number of stories that could be told; however, they open up other possible ways of knowing and being, and as a result it is now impossible for any of us to conceive of science as we once did.

Theoretical Context

We found a variety of methods and analytic tools important to this project. First and foremost is the method of memory-work developed by a group of German women who formed the *Frauenformen* (roughly, women's forms) collective in the late 1970s. They are an editorial group that formed within *Das Argument*, a Marxist journal originally launched in 1959, with the goal of bringing together their work in the Women's Movement and their work on the journal. Their reference point is Marxism, and their stated aim is to "inscribe feminism into the Marxist framework" (Haug, 1987, p. 23). Memory-work was developed for a writing and research project focused on women's socialization. The collective was dissatisfied with then-current theories of socialization coming out of the disciplines of sociology and psychology. They felt that both disciplines neglected discussions of girls' socialization, and when they did include girls, they were treated as objects of various socializing agents acting on them and forcing them to take on particular roles. The question of how individuals appropriate certain ways of behaving or how they learn to develop one set of needs as opposed to another was neglected (Haug, 1987). As Haug notes, these theories never address the "existential afflictions and obstacles" (p. 24) faced by girls as they grow into womanhood.

Haug's method of memory-work and its related assumptions about socialization share many aspects with social constructivist descriptions of development as expressed by Vygotsky (1978, 1986), Wertsch (1991), Rogoff (1990), and others. Constructivism refers broadly to the belief that the activity of mind participates in the construction of the known world. The person is involved in all knowing (Overton, 1998). This position is often contrasted with realism, in which the world is mind-independent. The perception of the world for the realist is direct rather than mediated by the mind.

The critique that Haug leveled at earlier theories can be applied to the theories of Vygotsky, Wertsch, and Rogoff as well. They do not specifically address girls' socialization, and while all of them theorize a more active individual, with agency or power, there is far greater research attention paid to the socializing forces of a particular culture than to the power of an individual to impact those forces. Despite these criticisms, we found some aspects of social constructivist theory helpful in our work.

Three aspects of Vygotsky's (1986) social constructivist position are important in our work. First, Vygotsky described the distinction between spontaneous and scientific concepts in the child's thought, a distinction we discuss in chapter 2, "Nature to Natural Science," and consider in relation to personal science in chapter 4, "Making Sense."

Second, Vygotsky's writing on imagination proved helpful in understanding our personal science. We comment in chapter 6, "Making New Meaning," on his ideas about the transformation of childhood play into scientific and artistic creativity for adults (Smolucha, 1992).

Third, we also found useful Vygotsky's concept of the zone of proximal development, characterized by a theoretical space between a child's ability to solve a problem with assistance and his or her ability to solve the same problem without assistance. In chapter 8, "The Power of Girls," we explore this concept and the related concept of guided participation (Rogoff, Mistry, Concu, & Mosier, 1993).

Haug's (1987) theory of memory-work takes social constructivism an important and critical step further than Vygotsky and his successors. Instead of merely describing how the child appropriates and is constructed by cultural and community values, memory-work asks us to scrutinize those values in ourselves. As participants in memory-work we analyze how those values have shaped us through memories of our socialization. Theory and experience are brought together, and in this merger, we see personally and collectively how we come to tell the particular stories we tell of our lives and of the world. Through the scrutiny that memory-work demands, we see how values that function to reproduce the dominant culture such as those related to racism, classism, sexism, heterosexism, and so forth, come to influence the stories we tell. Specifically in our study, we can experience firsthand how patriarchal values influence the stories we tell about nature and scientific ways of knowing the world. In this sense, memory-work goes well beyond the benign descriptions of social constructivism. In using the method, we attend to aspects of our lives that are oppressive and unjust and come to recognize how oppressive values operate in our lives, so that we may be positioned to undermine those values.

A theoretical perspective that was helpful both in analysis of the memories and in working through methodological issues was Cataldi's (1993) study of emotional depth and physical distance, in which she drew upon the work of Gibson and Merleau-Ponty. Depth, she suggests, is more than just the distance from one point to another. It may also encompass emotional depth, i.e., the "lived distance, which is a space of personal outlook or concern, the space in which I am 'bound' to the things I care about and in which living movements or 'e-motions' take place" (p. 53). This perspective on depth aided in examining our connections to nature and in analyzing our relationships with each other as we engaged in memory-work. For example, one kind of lived distance in the academic world entails fear of being revealed as not knowing enough, which may lead us to isolate ourselves from our colleagues. For us, this distance initially functioned as a barrier to understanding each other's stories. Additionally, we are social and natural scientists and therefore supposed to distance ourselves from the objects of our study. However, memory-work insists on the inseparability of subject and object in that it rejects the privileging of distant impersonal knowing. Cataldi's (1993) treatment of depth led us to expect that the distance between subject and object would shift, sometimes diminishing, sometimes increasing. We regarded this as a dynamic depth as we grew accustomed to the possibility of the unity of subject and object.

We also critique those values emerging from the positivist position of elite science. This science includes the process of investigation as well as the body of knowledge provided by a hypothetical-deductive approach for the observation of natural phenomena. This characterization of scientific knowledge is mirrored in the lay perception of science and hence permeates Western dominant culture. Further, as Keller (1985) points out, the male founders of Western science clearly depended on the language of gender in establishing a science based upon a "masculine" philosophy. The values that emanate from this view have consequences for our sense of what counts as knowledge in the world and who may generate that knowledge. Relevant to our study is the impact on women scientists who may have had to move away from their identification as feminine to take part in masculine control of nature or have had to redefine terms as neither feminine nor masculine in the conventional understanding of meaning.

Memory-work has led us to think differently about our relationship to nature, what counts as knowledge, and our own lives as scientists. Our investigation is focused on nature, presuming that understanding how we have been socialized to think about nature would illuminate our journey into science.

As we talked about this work with colleagues and presented the results of our research to professional audiences, we have encountered considerable curiosity and excitement about the methods. Colleagues talked with us about how they might use some variation of the method in their classrooms or workshops. Others found that our journey and findings stimulated their thinking about their own science, their paths from their experience in nature to work in science or to turn away from work in science. We began to consider a book rather than just a series of papers to report our work and answer some of the recurring questions.

Organization of the Book

In the course of analysis, some memories, such as the sandbox memory, came up in discussion over and over again. They stimulated our thinking and pushed us to use different lenses for seeing our socialization in relation to nature and science. You will find repetitions of parts of key memories throughout this book because we found that looking at them in different ways led us to the themes that emerged from our data. As a result, all of the memories were coded repeatedly and separately for evidence of the sensuous, for ways in which the classical elements and their transformations appeared in memories, for evidence of play and experimentation, for appearances of family members, to name several of the categories. When the categories that resulted from coding emerged, we had the first traces of the themes around which the following chapters revolve. With the themes conceptualized, we paired off to draft the first versions of chapters. Each chapter has evolved from the original work of two or three members of the collective. We revised and reworked each of the chapters in response to discussion by the entire group.

Our work is divided into two broad sections. In the first, chapters 2 and 3, we present our theoretical locations and the methodology of memory-work. In the second section, chapters 4 through 9, we describe the themes that have emerged from our analysis of our memories.

In chapter 2, "Nature to Natural Science," we discuss the nature of science and its relation to the natural world, and we introduce the problematic position of women in science. Building on previous research about children and science and children and nature, we examine multiple views of science focusing specifically on women's ways of knowing, feminisit critiques of science, and our interrelationships with nature and science. In chapter 3, "Memory and Memory-Work," we detail the methodology of memory-work and the view of memory that underlies this research project, specifically. We discuss how our memories reveal who we are in the present in contrast to who we were in the past.

Our findings comprise the second section, and they are organized according to the themes that emerged from this multiyear project: the sensuous, metaphor, creativity, family landscapes, and power.

In chapter 4, "Making Sense," we discuss the sensuous in our memories in relation to how the senses meld with the social in influencing our connections to nature. As a result of memory-work we discovered the importance of the sensuous and reclaimed our pleasure in the natural world. Here, we focus on the ties between sensuous experience and the emergence of a personal science, which is embodied during everyday activity.

In chapter 5, "Metaphor: Girls in Their Elements," we examine the metaphors in our memories as an approach to understanding our socialization in nature. We also consider how the natural elements contribute to our metaphors for fear, success, relationship, growth, and development. We found that some metaphors draw us in close connection to nature and some push us away from it. Our consideration of metaphor shows how Western culture functions to distance us from nature. We discuss the implications of this finding for our interactions with science in schools and traditional science.

In chapter 6, "Making New Meaning: Creative Acts," our analysis is focused on the creativity in our memories of our interactions with nature. Creativity, by our definition, is making new meaning, and we made new meaning for ourselves in nature when we played, solved problems, imagined, explored, and experimented. We explore the meanings we created within the physical and social boundaries of creative spaces. By focusing on these boundaries in our memories we learned how creativity is hindered or facilitated in our lives.

In chapter 7, "Family Landscapes in Nature," we examine the ways in which our families are and are not portrayed in our childhood and adolescent memories. This analysis revealed much in terms of how our relationships to the natural world are shaped by our interactions with parents, siblings, and friends. The relationships and related emotions described in the memories signify what is valued and hence what we learned from our families about the natural world.

The last theme, focusing on power, is discussed in chapter 8, "The Power of Girls." Power is present in all of our memories. As young girls we felt powerful in nature and we were able to act powerfully in nature. Powerful women were present in some of our memories, and they facilitated our activities in nature. Based on our early memories, we came to recognize that nature could be powerful and we sometimes lost power when adults intervened. The quality of power changes from our early to our later memories and, in part, appears to reflect how we gradually learned to distance ourselves from nature.

In the concluding chapter, entitled "Interruptions," we explore the interruptions in our views of self and the world that have occurred for us as a result of memory-work. Our initial view of the stories about our lives in the natural world focused on narratives that reflected dominant and sometimes invisible Western cultural values. Through memory-work, we named those values and excavated stories that interrupted the dominant cultural narrative of the natural world. These excavated stories show that while the past can exert an influence on the present, particularly in the case of childhood trauma, the present exerts an enormous influence in defining the past. Who we are now at this moment determines the kinds of stories we tell about the past. Recognizing this means that the past can be narrated as many possible stories, and this, in turn, opens the present to other possibilities. We can, for example, rethink and actively respond to our distanced and, hence, often destructive relationship to nature. We explore some of these possibilities when we consider how our relationship to nature has been altered as a result of this research. Finally, we discuss how we have been transformed through memory-work and how these transformations have worked their way into our teaching and research.

> No one ever told us we had
> to study our lives,
> make of our lives a study, as if learning
> natural history
> or music, that we should begin
> with the simple exercises first
> and slowly go on trying

the hard ones, practicing till strength
and accuracy became one with the daring
to leap into transcendence, take the chance
of breaking down in the wild arpeggio
or faulting the full sentence of the fugue.

from "Transcendental Etude"
  The Fact of a Doorframe: Poems Selected and New (1950–1984)
Adrienne Rich, 1984

Guide to the Reader

You may be an educator, a scientist, a feminist, or simply a curious reader. We think that readers who have different interests may wish to approach this book in a variety of ways. Some readers may wish to skip over chapters 2 and 3 on theory and method and complete them after they have read the findings and conclusions of the study in the remaining chapters.

The reader will find helpful two appendices. Appendix A is a complete list of the memories we generated during the course of this study. Appendix B is a directory of all memories quoted in the text, organized by location in the book, the cue that stimulated the memory, the author, and the period in life (childhood, adolescence, or adult) from which the memory is drawn. Both of these resources immediately follow "Illustrations."

Each extensively quoted memory is preceded by the author's pseudonym, the period from which it is drawn, the cue for the memory, and the name of the memory. For example, the memory that opens this chapter is preceded by Bell, Adolescent, Fire~Sweaty Palms Memory. Quotation of memories and memory fragments appear in italics. Occasionally, we quote directly from transcripts of our research meetings. The date of the research meeting and transcript page follow the quotation. Not all of our memories were used or presented in this book, and sometimes when memories are used in the text, they are incomplete. In this way, we reduced the redundancy of repeated readings.

PART ONE

CHAPTER 2

Nature to Natural Science

"Sikrit stuff for sikrut envinshuns-sikrit" (Sign on a laboratory door warning of secret inventions, drawn by a third-grader).
—Chambers, 1983, p. 264

Our interest in how women's relationship to nature develops and how it is
linked to our evolution as scientists drives our research as a memory-work
group. However, the connections between nature and science are neither
linear nor easily teased out. Our socialization—as girls, women, scientists, as part of
nature, and observers of nature—is integral to our views and is reflected in them.
Popular conceptions of science and the masculine character of science, to name just
two social forces, affect adults' and children's notions of what science is like and
who may be a scientist.

In this chapter we emphasize the character of science. In later chapters we will
emphasize this research group's experiences in nature and how they demonstrate
our early interest in the natural world, our scientific interactions in it, and our
largely contrasting experience with science in school. Thus, this chapter frames cultural notions of science and scientists as a backdrop for the following chapters in
which we present our analysis of our experiences in nature and their connections to
the science we now see differently as a result of memory-work.

Here, we first explore the literature that suggests that the masculinist context of
science, historically and today, affects children's and women's views of science in
ways that may discourage their interest and participation. We begin with children's
first impressions of science. As a result of our research, we believe that we must be
more attentive to children's perceptions of science and their experience in nature if
we are to encourage their exploration of the natural world and science. We then examine some conventional conceptions of science because these affect what we commonly believe counts as science and who counts as a scientist. These include some

of the ideas that we, the authors, once believed constituted the essence of science. The feminist critiques of science that follow indicate several problems that masculinist science presents to the thinking and practice of science by women. Women's responses to the current climate of science and the challenges that women encounter in becoming scientists are discussed next. In the final section of this chapter we return to children and the importance of play in exploring nature and reinforcing the scientific efforts of children.

Children's Perceptions of Science

Although most children know very little about formal science, their perceptions of scientists are useful to understanding preconceptions of who can be one. When young children were asked to draw a scientist, Chambers (1983) found seven elements that served as indicators for the standard image of a scientist. These included a lab coat, eyeglasses, facial hair, lab equipment, books or filing cabinets, technology, or relevant captions such as formulae or "eureka." By the second grade, children incorporated at least two of these features and by the fifth grade, three or four of them, rather like adults who included four or five of the features in their drawings. On the basis of such drawings, a description of the stereotypical scientist might be a bearded man with glasses who wears a white coat and works in a laboratory. His work space is filled with equipment such as flasks and Bunsen burners. The scientist writes neatly in black notebooks, reads a book, or shouts "I have found it!" The statement quoted at the beginning of this chapter also emphasizes the "secret stuff" of science, a child's perception of the closeted nature of science.

The young children studied by Jarvis (1996) also drew different images of scientists as they grew older. She found that children younger than 6 years old typically drew what they wanted to (e.g., a house) rather than a recognizable scientist. Between 6 and 8, they were likely to depict either a generalized individual, that is, without gender or occupation indicated, or they provided drawings of teachers, artists, or stereotypic male scientists. After age 8, they tended to draw white men wearing glasses and lab coats. Lutz (1994) solicited drawings and descriptions of scientists by sixth graders, which were reported anecdotally in connection with reviews of science books for children. Among the definitions of a scientist accompanying those drawings, most referred to the scientists as male. One of Lutz's subjects stated, "A scientist is weird—probably never sleeps or eats because he's always working." Another, however, commented, "I don't think scientists are all people with funny hairdos" (p. 575).

Stereotypical views of scientists were prevalent, almost universal, by age eleven among children in Jarvis's study. Rarely did children draw women or people of color as scientists. Interestingly, whereas discrimination on the grounds of race was considered unjust by these children, there was persistently more disagreement about the importance of a woman's occupation, with many of the boys feeling that mothers should be at home with children, rather than following their careers as scientists.

Young children in Jarvis's study did not view activities as science except when they were told explicitly that those activities were science. After exposure to accounts of science or science experiments, these children recognized scientific activities that were not labeled for them as science. The 5- and 6-year-olds who drew "recognizable" scientists and drew themselves engaged in activities that might be construed as scientific had previous experiences in a "Science Room," i.e., they internalized someone else's designation of what is scientific. In fact, most people, perhaps even most scientists, depend at some point in their development on someone else's notion to frame for themselves what is scientific and what is not. What are the conventional views of science?

Views of Science

Many scientists and philosophers would agree that science is a systematic way of coming to know and that the knowledge that results is highly valued in Western culture. It is important to remember that science represents only one way of understanding the world; everyday experience, poetic insight, and religious revelations are obvious examples of others (Baltes, Reese, & Nesselroade, 1977). Historically, the roots of science lie in methods for describing and controlling the material world. Definitions typically focus on observation, reliability, and usefulness in generalization and explanation. However, as Mayr (1997) pointed out, agreement about a definition of science is difficult because "science is both an activity (that which scientists do) and a body of knowledge (that which scientists know)" (p. 25). Both the activity and the knowledge change over time, and their limits are frequently debated among working scientists. Indeed, the American Physical Society has been debating the very definition of science for the past several years (Holden, 1998). Macilwain (1998) reported that, at the time, the debates about definition centered on three elements: that science is based upon empiricism, that reproduction and verification are important, and that constant questioning is important as a self-correcting mechanism (Holden, 1998). Needless to say, individual scientists may also hold idiosyncratic notions of science. James Watson, Nobel Prize-winning molecular biologist, has been quoted as saying, "There is only one science, physics: everything else is social work" (Rose, 1998, p. 8).

Neumann (1997), among others, has noted that science may be defined differently across the various disciplines. Most people, he argues, think of physics, chemistry, or biology when they hear the word science. This might be the case, for example, upon hearing Byrne's (1993) argument that science is usually perceived as an intellectually based curiosity that leads to knowledge of principles and a set of abstract attempts to explain the world. Many authors tie this sort of definition directly to the natural world. For instance, Graziano and Raulin (1997) noted that the essence of modern science is the way of thinking, or asking questions to understand "natural events" (p. 2) with the eventual goal of understanding the natural universe. We will return to this point later in this chapter.

When "science" is mentioned, the natural sciences, particularly the physical sciences, are what people typically have in mind. In fact most people, including many scientists, do not know that the social sciences are part of the U.S. National Science Foundation. We raise this point because when we speak of the sciences in this book, we include the social as well as the biological and physical sciences. Not surprisingly, within the social sciences, definitions of science also vary depending upon the theoretical framework in use. Many, but not all, are based upon natural-science models, in the sense that they typically focus only on observable events, and empiricism is emphasized.

The positivist tradition separates empirical science from nonscience, and this view has dominated Western science for at least 100 years. Overton (1998) characterized positivism as employing two criteria in descriptions and explanations of phenomena: First, a proposition must be reduced to words whose meaning can be directly observed; second, any generalization must be directly induced from "pristine observations" of the scientist (p. 158). For example, we might wonder whether plants need light to live. We might propose that healthy plants shut away in the basement turn pale and die because they do not receive enough sunlight and that beans, planted in the deep shade of a large tree, grow for a while and then also turn pale and die for lack of sunlight. We might, as scientists, experimentally explore the effects of sunlight on plants. Upon observing the results, we conclude that the poor health of our shaded plants is the result of no sunlight. Therefore, sunlight is needed by bean plants for health, a generalization based on our controlled experimental observations ("pristine observations"). Logic, according to Overton, was brought into the system to keep the science value free. He claims that logical positivism is currently in decline, but this approach, recognizable as science's hypotheses and deductions, still permeates views of science. The criterion of value-free science is closely related to the notion of objectivity, which is commonly understood to mean dealing with phenomena without the distortion of personal feelings or interpretations. This distance that we typically associate with the objectivity of empirical science marks elite science today, that is, the work of prestigious nationally and internationally recognized laboratories. However, experimentation itself does not always entail this distance. We use experimentation every day in finding new ways to do all kinds of tasks, as we will describe from the experimental approaches of girls in later chapters.

At the same time, successful working scientists are by no means limited to the distance often associated with the scientific method. A *Feeling for the Organism*, Keller's 1983 account of Barbara McClintock's path toward understanding the interaction of genes and their environment, describes an approach to science that seems on first glance to be highly unusual, perhaps because it rarely has been acknowledged previously. McClintock's work in corn genetics, for which she received the Nobel Prize, has come to exemplify the usefulness of relational thinking. As Keller has pointed out, this kind of thinking, which is regarded by some as feminine, does not define women's approaches to science (e.g., Keller as quoted by Barinaga, 1993; Tobias, 1992). Many observers of women in science argue that the flexibility to combine alternative approaches with traditionally objective ones is not seen in all women, and it is certainly characteristic of some men as well. Numerous scientists, male and female, do what Keller said of McClintock, "She made use of the full range of human capacity . . . and all her intuitive strengths, in the service of science" (p. 392). Kuhn (1962) argued that the shared commitments to concepts, theories, instruments, and methods define the paradigm and research tradition of a discipline. Many social scientists find untenable a science that is restricted to hypothetical-deductive knowledge based on the observed. In fact, many accepted "facts" of other areas of science are not based on direct observation at all, for example, neither the quarks of physics nor the double bonds of chemistry have been directly observed. The researcher interested in cognitive processes in infants may find the relational approach more appropriate in research, in which interpretation, reason, and observation are equal partners in the scientific enterprise. Once we acknowledge that "all data are theory driven," interpretation's role in science becomes very clear (Overton, 1998). For some researchers (such as us) the relationship between subject and object, as well as interpretation of observations and theory, is key. Nonempirical approaches are admissible in this kind of science.

Feminist Critiques of Science

Typically we are introduced to positivist science in secondary school through the scientific method, and science and the method become almost indistinguishable. As feminist critics of science have said, "the scientific method is supposed to be powerful enough to eliminate any social biases that might find their way into scientific hypotheses because of the social identity of the scientist" (Harding, 1992, p. 60). Nevertheless, feminist critics note that science reflects the concerns of dominant social groups, such as white middle-class men; social bias is inevitable. In addition, science often ignores the issues of concern to women and other underrepresented groups in its work. For example, Campbell and Schram (1995) analyzed the research methods textbooks in psychology and the general social sciences, which are primary tools in training future researchers. They concluded that feminist challenges to be nonsexist and profeminist in such texts were met only minimally by nonsexist language in these books. Very little discussion of feminist critiques of research issues was found among them.

Eichler (1988) maintained that androcentric approaches to social scientific research are common and pointed to the need to recognize and address sexism in conceptualizations of research design, methods, interpretation, and policy recommendations. Some feminists have also seen the preoccupation with quantitative analyses in research as evidence of a historically masculine focus on abstraction and distance in analysis and interpretation of data. In many quarters, calls for the integration of both qualitative and quantitative approaches to research questions (e.g., by Jayaratne & Stewart, 1991) receive little or no response. The steadfast reliance solely on quantitative approaches continues to be common in the natural sciences and to a lesser extent in the social sciences.

More generally, Bryman (1988) described five assumptions of positivism that are at odds with many contemporary views of science and that many women may find problematic:

1. transfer of methods from the natural to social sciences is appropriate;

2. only observable phenomena can be subjected to research;

3. knowledge is an accumulation of verifiable facts;

4. scientific theories provide the necessary foundation for empirical research;

5. a scientist must take a value-free position in relation to the research.

Such assumptions may well be objectionable to many who doubt it is possible to take a value-free position in research or who see the analytic needs of the social sciences as different from those of the physical sciences, for example.

Deficiencies of the culture of science, it has been argued, may deter women from pursuing science as a career (Maynard, 1997). Keller (1985) forcefully described the explicit reliance on the language of gender among the male founders of Western science. "They sought a philosophy that deserved to be called 'masculine,' that could be distinguished from its ineffective predecessors by its 'virile' power, its capacity to bind Nature to man's service and make her his slave (Bacon)" (p. 7). Though the language of science is rarely so explicitly gendered today, the longstanding link between nature and the feminine and science's power over nature remain, however veiled they may be (Merchant, 1980). As a result, women may feel constrained in dealing with the implications of those assumptions of power over nature and nature as feminine (Keller, 1985). Some women scientists repress their feminine identities and, for example, become "one of the boys" in order to share in masculine mastery. Other women scientists respond to the gendered language of science by, for example, suggesting that a formerly masculine-identified process, such as formation of the pilus in bacterial conjugation, is neither active nor passive, masculine nor feminine (Spanier, 1995). Masculine discourse is embedded in our science to such an extent that it often escapes our notice.

Women's Views of Science

Barr and Birke (1998) contend that many women share a view of science that they encountered in their research, namely that "science is everywhere, yet it has nothing to do with me" (p. 1). As we have indicated in chapter 1, "Introduction," the five of us, too, sometimes have felt alienated from science. Feminists often have suggested that people who feel that they operate at the margins of some part of a culture may be placed ideally to critique and change that culture. We would suggest that girls and women may have the potential to alter masculine discourse in nature and science.

In *Women's Ways of Knowing,* Belenky, Clinchy, Goldberger, and Tarule (1986) explored the kinds of "knowing" that women often use. A common approach, they maintained, is to put one's self in the shoes of another in order to understand that person's perspective and ideas. This approach runs counter to objectivity, which is commonly believed to be necessary to science. In fact, the relational frame in which many women think and work creates some very different views of science. As a participant in Belenky et al.'s study remarked, "Science is a moral art, dictated by the human heart and mind. It was subjective and is subjective. Science is a creative evaluation of facts, of demonstratable [*sic*] happenings" (1986, p. 138).

What other views of science do we see among those who are not traditionally considered scientists? A farmer who lives in the same area for 60 years comes to predict the weather, or the course of a season, fairly accurately. That is often called common sense. When the climatologist, who works with observations from remote atmospheric sensors, creates graphs and databases on a computer and makes the same predictions, it is called science. Given science's emphasis on distance, it is not surprising that what we experience personally comes to be common sense and what the distant experts know is thought of as science.

Nonacademic women may perceive science as irrelevant, as Barr and Birke (1998) suggested, in part because these women see science as part of the "master narrative" of the dominant culture in which they are marginalized (Lyotard, 1984). For example, Afro-Caribbean women who had recently immigrated to Great Britain observed to Barr and Birke that science was not part of their original culture and that science denied the validity of understandings they had learned in their own cultures. These women drew a distinction between academic science and common sense. Science is a set of explanations they claimed not to know or understand, and common sense refers to the knowledge they own and acknowledge. As Wynne (1991) has suggested, assimilation of scientific insights by the public is not so much based upon intellectual capacity as it is on "social access, trust and negotiation, as opposed to imposed authority" (p. 116).

Barr and Birke (1998) maintain that "the power to name what counts as science does not belong to women" (p. 34). It is those in elite science who constantly police the borders, negotiating what counts and what is good or even valid. At the same time, women are pointing out other kinds of science, for example, the science of women's kitchens and gardens (Hubbard, 1988) and the personal science that we see evident in our memories, to mention only two. In later chapters we will explore how our use of memory-work has led us to redefine science in terms of our own experience and how distant these definitions are from the traditional definitions of other scientists.

Becoming a Scientist: Conflicts in Identity

The number of women in the physical or natural sciences and engineering increased from 1960 to 1990; 19,362 bachelor's degrees and 381 doctorates in these areas were earned by women in 1960, and in 1990 the comparable numbers were 123,793 and 6,274 (Barber, 1995). However, the proportion of women receiving bachelor's degrees in these areas who went on to earn doctorates in the natural sciences and engineering did not change from 1970 to 1990. Regardless of the gains women have made in the past, science still seems to be an inhospitable place for women and girls (Didion, 1993).

Byrne (1993) addressed the underrepresentation of women in the sciences in terms of the Snark Syndrome, quoting Lewis Carroll who wrote, "Just the place for a Snark! I have said it thrice: What I tell you three times is true" (Dodgson, 1966, p. 5). Relying upon constant repetition, many have asserted that male and female minds are different and that girls are innately mathematically deficient as compared to boys. Girls' interests are patently different from the interests of boys, so girls do not enroll in natural or physical science, their assertions continue. Similar is the Snark-like repeated assertion that because we now have equal education for boys and girls, underrepresentation will just disappear over time. Byrne examined these assertions and separated the policies emanating from them from the empirical data available. She acknowledged progress encouraging girls in science while proposing that we make science more welcoming to women and provide more role models. Math, she believes, is critical in influencing girls' views of science. She suggests that we consider instituting more single-sex mathematics education, since girls are more successful in these contexts.

More than a decade earlier, Fausto-Sterling (1997) identified other problems in the classroom that discouraged girls from entering science: the invisibility of women in curricular material, the hierarchical nature of traditional science, and the competitive nature of the science classroom. Since that time, others have reiterated these concerns about competition, isolation, and the lack of female role models, as well as citing the lack of value accorded to marriage and family as issues that impact women's participation in science (Civian, Rayman, Brett, & Baldwin, 1997; Seymour & Hewitt, 1997; Tobias, 1992).

Tobias (1992) argued that the data on the participation of women in science could be interpreted to demonstrate that women do not have the staying power or scientific intelligence to stick with science or that the culture of science does not know how to appeal to and nurture women's talent. Tobias described one of her own students who concluded that while her IQ was sufficient to do physics, her OQ (obedience quotient) was too low. The student wanted to interact creatively with the material and interpret the findings.

Why do students remain in or leave science majors as undergraduates? Seymour and Hewitt (1997) analyzed interviews with 335 men and women students at seven universities, examining students who remained natural science, math, and engineering majors and those who switched out of them to other majors. The prime reason for leaving science, given by both men and women, was that they had lost interest, and about 90% of both men and women who switched cited poor teaching by science faculty as a concern. This finding suggests that more students might stay in science majors if excellent teaching that sparks interest were available. These researchers also reported that women seem to lose confidence in their ability to excel at these sciences as undergraduates to a far greater degree than men do, even when they are doing relatively well in terms of grades. Women often turn to faculty mentors for reality checks on their performance, and those who do not receive the information they need may turn to other avenues of support and verification or, alternatively, leave these sciences. This pattern suggests that responsive mentors may help women in the sciences realistically assess their performance and thereby provide an important kind of support needed by the women who are performing well but are not certain that their performances are good enough.

Didion (1993) suggested that some academic environments, in fact, are more hospitable than others for women entering the natural science and engineering fields. Women receiving doctorates in science and engineering were less likely to have earned their baccalaureate degrees at research-intensive or doctoral institutions than at a liberal arts institution, according to a National Science Foundation (NSF) special report (Hill, 1992). Therefore, Didion suggests that in the larger and more impersonal research institutions, having senior professors who are more accessible in introductory courses and more encouraging of women in the sciences would be helpful.

The decline of self-confidence and what Leslie, McClure, and Oaxaca (1998) term self-efficacy begins long before college. Rita Colwell, Director of the National Science Foundation, refers to the "'valley of death' in education, when girls in grades 4 through 8 are in subtle and not-so-subtle ways discouraged from pursuing science and engineering" (Colwell, 1999, p. x). The American Association of University Women (AAUW, 1992) reports that girls begin to lose confidence and interest in math and the sciences in adolescence. Eisenhart and Finkel (1998), interpreting this shift in interest, suggested that young people in the United States first face the demands of science in high school or college, where they find choices between "greedy" inflexible programs and more flexible courses, degree programs, and possible occupational options. "Greedy" refers to large numbers of courses that demand many hours in the laboratory and in preparation for class; in combination with rigid and narrow program requirements for specific courses, which must be taken in a particular order, little flexibility results. At this point, men and women may make different choices influenced by different sets of cultural and social expectations. Eisenhart and Finkel suggest that many women move toward flexibility, partly in response to their coming to terms with the social expectations for women. They may take on the commitment of time and energy necessary to develop the identities associated with traditional science and limit their social involvement outside of science activities. Alternatively, they may opt for flexibility in attempting to juggle the high demands of traditional science with those of multiple social roles according to Eisenhart and Finkel. We would add that many elect a third approach, pursuing the social sciences.

In addition to environmental constraints, women are concerned about having children and the dilemma of their timing (Tobias, 1992). Seymour and Hewitt (1997) reported the concerns that women students in the natural sciences and engineering voice regarding the difficulties of combining careers and parenthood. As a senior in that study said: "If you have the time, you don't have the money. If I have kids, it won't be for a while. I don't want to have them when I'm 40. I dunno. Maybe I'll skip the kids and have grandkids. I haven't found a way to do that though!" (p. 292).

The Wellesley College Pathways in Science Project (Civian et al., 1997) followed women with bachelor's degrees in the natural sciences for 25 years beyond graduation. Women in science professions were less likely than those in other professions to have children. Furthermore, once a woman on the natural science track had a child, she was at greater risk of dropping out of science at all points along the pipeline than were those who were not mothers. Women in the natural sciences were also much less likely than women in other professions to say that their occupations were compatible with raising families.

The dilemma of whether to become a parent often surfaces after natural scientists finish the doctoral degree and complete postdoctoral studies. In Sonnert and

Holton's (1995a, 1995b) study of the careers of former NSF postdoctoral fellows, nearly all the women with children reported that being a parent had influenced their scientific careers, and two-thirds of the men made similar responses (1995a, p. 159). Although at first glance it seems that the disadvantages of marriage and parenthood for women in a system that does not support child rearing should be clear, the balance of advantage and disadvantage is complex. So, for example, rather than detracting from their careers, women (and men) reported that children provided major emotional satisfaction that put their careers in perspective. Sonnert and Holton urge us to look at marriage and parenthood as presenting both opportunities and problems. "Women scientists are faced with the dilemma of 'synchronizing' the often conflicting demands of three clocks: their biological clock, their career clock (such as their tenure clock), and their spouses' career clock" (1995a, p. 161), particularly because women scientists are far more likely than men scientists to marry partners who also have doctorates (62% versus 19%).

Evetts's (1996) study of scientists and engineers in industrial organizations re-emphasized that increases in work responsibility and time commitment are expected parts of career development for men. Although work has become an important part of women's identities, they (much more so than men) must integrate the pressures of career development with greater responsibilities associated with family and other social roles. However, as Bateson in *Composing a Life* (1989) reminds us, women have always dealt with multiple tasks and responsibilities, often simultaneously, in their everyday lives. Perhaps here too we, like Sonnert and Holton, might see opportunities as well as problems for women balancing their lives as scientists and members of families. Bateson takes women's lives as a model for thinking about the potential for transferring learning and asks "What insights arise from the experience of multiplicity and ambiguity?" (p. 10). We might suggest that such experience may contribute to a scientist's willingness to hold several theoretical or experimental views open in constructing scientific generalization or explanation. It might also facilitate flexibility in combining the contributions of the members of a research or teaching group or in coordinating the work of people with unconventional schedules.

Examining role ambiguity from a different angle led Eisenhart and Finkel (1998) to question traditional ideas about where science is pursued, what it looks like, and how it is learned. They looked at women practicing science at the margins, that is, not in the laboratory settings of elite science. Several women in their study, for example, were learning science and finding professional success in a nonprofit corporation that worked to protect biodiversity. Priorities and organizational patterns in such places are different from those of elite science in a number of ways. Applied rather than experimental science may be emphasized; the relevant and contingent tend to be more important than abstractions and the unvarying; and the whole enterprise is practiced in "the messiness of public activity rather than in the more private and controlled spaces of classrooms or laboratories" (p. 232). Eisenhart and Finkel contend that women may find these features more interesting and perhaps more comfortable because they fit well with their experience, with the social roles and cultural expectations of women's lives. As these women have learned to handle the multiple lower-status priorities in such a setting, they also have "learned to prize their distance from the priorities of elite science" (p. 232). They have opted for a different kind of science.

Children in Nature

Our experiences in nature as explored through memory-work suggested to us, too, ways of thinking differently about science. Because many of our findings in this study came out of analysis of our childhood memories, we look once more at children here. It is important to think about children in nature to remember and understand the changes that girls experience in their relationships to nature as they grow up to become adults.

Biologist Neil Campbell said that science as a way of knowing "emerges from our curiosity about ourselves, the world, and the universe. . . . At the heart of science are people asking questions about nature and believing that those questions are answerable" (1993, p. 15). Although children have limited experience of science structured by teachers or parents early on, most have some contact with empty lots, flowerpots, insects, or pets. Vygotsky (1986) referred to the concepts that grow out of such experiences as "spontaneous" in contrast to the formal scientific concepts, which are developed using cultural tools. Children reflect, think about thinking, and have multiple relationships with nature. Furthermore, children often have experiences in these settings that are exploratory or investigative, though they are more likely to be labeled play or social interaction than science.

Play spaces are important arenas for exploration, investigation, imaginative play, and social interactions. We might expect that significant experiences in the natural world take place in play spaces and may well have connections with the development of skills and interests that later are labeled science. As Emily Barnes, a child cited by Lutz (1994, p. 575), said, "A scientist is a person who . . . is very observant. A scientist is an inventor, a teacher, or just a child, playing in his backyard. You and I are scientists. A scientist can be anyone." We believe, as a result of memory-work research, that she is correct. In the following chapters, you will see how this method provided the five of us an opportunity to explore our socialization in nature and our fundamental beliefs about science. We chose cues from the natural world in this research, presuming that these would elicit memories that would illuminate our journey into science. The following chapters recount the method and our insights from its use.

CHAPTER 3

Memory and Memory-Work

Since women have been excluded from the creation of formalized knowledge, to include women means more than just adding women into existing knowledge or making them new objects of knowledge. . . . Including women refers to the complex process of redefining knowledge by making women's experiences a primary subject for knowledge, conceptualizing women as active agents in the creation of knowledge, including women's perspectives on knowledge, looking at gender as fundamental to the articulation of knowledge in Western thought, and seeing women's and men's experiences in relation to the sex/gender system.

—Anderson, 1990, p. 120

Memory-work as a social-scientific methodology challenges a number of conventions associated with the generation of scientific knowledge. It challenges the schism between theory and everyday experience and validates personal experience as a legitimate source of knowledge. Therefore, everything about our experience is open to interpretation. Nothing is accepted as fixed, such as character traits or "typical" ways of behaving. Instead, there is a search for the ways in which we actively participate in shaping our pasts (Haug, 1987).

The traditional distinction between subject and object is challenged with the use of memory-work. Both subject and object implicate "fixed and knowable entities" (Haug, 1987, p. 35), and these ideas conflict with memory-work's aims of change and potential liberation. Haug cautions that memory-work is only possible if individuals assume the roles of both subject and object.

The feminist-based memory-work research projects of Haug (1987) and Crawford et al. (1992) guided our efforts. Haug's text "records a collective's attempts to analyze women's socialization by writing stories out of their own personal memories: stories within which socialization comes to appear as a process of sexualization of the female body" (p. 13). The collective used the method termed "memory-work," derived from Haug's theory of socialization, "as a bridge to span the gap between 'theory' and 'experience'" (1987, p. 14).

Haug's collective first used personal memories related to love, marriage, happiness, and the desire for children to study various forms of feminine socialization. They also studied various parts and aspects of the female body, such as legs or body hair, and how these parts of the body come to be sexualized. Very generally, the method involves analyzing socialization by "writing stories out of their own personal memories" (Haug, 1987, p. 13). The eventual aim, for Haug, is liberation from oppressive social structures through understanding the process by which we are socialized and how we participate in our socialization.

Intrigued by a theory and method that requires collective participation and the collapse of subject and object (i.e., the researcher as her own subject), the Australian collective (Crawford et al., 1992) selected emotion as a topic. They used memories cued by terms for emotions such as sadness or happiness to examine the ways in which women socially construct themselves. Their perspectives on the purposes of research seemed ideally suited to our own work.

> Memory-work allows the investigation of processes, which involve the social construction of selves, and is not individualistic. Working at the interface between the individual and society, looking at ways in which individuals construct themselves involves doing research in an area, which both sociology and psychology might claim, but for which neither provides fully adequate tools. (Crawford et al., 1992, p. 4)

Who We Are

We are white and middle-class, lesbian and heterosexual women. Some of us have partners; all have children, and we range in age from 45 to 61. We have differing religious affiliations, and we have lived throughout the United States. Our academic disciplines include biology, child development, special education, educational psychology and educational administration. Our life experiences, though similar in terms of privilege, in many ways are quite different and reflect family traditions and understandings foreign to others in the group. Our mothers did everything from waiting tables to farming, teaching music to selling real estate, painting to running an office, teaching swimming to writing. Our fathers worked as doctor, military officer, engineer, heavy equipment operator and farmer, gas station owner, and truck driver. Patti grew up as the younger daughter in a small Midwestern town; Adrienne was an only child, spending some of her growing-up years in Washington, D.C.; Margaret was the youngest child in a family that moved from the suburbs of New York City to an Ohio steel town; Diane was raised in a Scandinavian farming community in the upper Midwest, the oldest of four girls; and Judy was the younger of two children and grew up in central Massachusetts.

We began this project at Oklahoma State University and met, during the academic year, on average once a week for two hours from January 1994 through May 1996. In 1996, Judy moved to Hofstra University. Thereafter, we met regularly online and intermittently for week-long research meetings and writing sessions through 2001.

Method

Using most of the injunctions developed by Haug's collective (1987), Crawford et al. (1992) modified and refined the method over the four years in which they studied emotion. They describe at least three primary phases of memory-work. The first involves the writing and collection of memories according to a set of specific rules. Each member writes her earliest memory of a particular episode, action, or event. The memory is written in the third person with as much detail as possible, including trivial and sensory components. At all times, interpretation, explanation, or biography should be avoided (Crawford et al., 1992).

The second phase involves collective discussion and analysis of the memories. Each member expresses opinions and ideas about each memory in turn. All should consider differences and commonalities or continuous elements across the memories. Autobiography and biography are avoided because they shift the focus from an analysis of social meanings to an analysis of the individual. It is important to use this approach because the focus on social meaning sets memory-work apart from therapy. It involves understanding the social construction of self, and the meaning generated moves well beyond the individual. As Haug (1987) notes:

> Human beings produce their lives collectively. It is within the domain of collective production that individual experience becomes possible. If therefore a given experience is possible, it is also subject to universalization. What we perceive as "personal" ways of adapting to the social are also potentially generalizable modes of appropriation. (p. 44)

Schratz and Walker (1995) also distinguish between memory-work and therapy. Memory-work, they argue, is intended to "close the gaps between theory and experience in ways that are intended to change the nature of experience, not simply to accept it" (p. 41). For example, therapy is sometimes meant to help an individual change so she or he can cope with or adjust to some set of social demands—the social is accepted as a given. In memory-work, the social is scrutinized and critiqued, and the insights gained from understanding how members have appropriated social structures can be used to loosen the restraints of those structures. This is memory-work's connection with political and social action.

Schratz and Walker (1995) point out that the tension in memory-work arises in the duality between the role memories play in socialization and social control and the potential those same memories have, through using memory-work, to undermine that control. In relation to our own group, we understand the development of our relation to the natural world differently for doing memory-work. We understand the socialization issues more clearly and can rethink them.

In addition to the activities already mentioned for the second phase of memory-work, members also identify clichés, generalizations, contradictions, cultural imperatives, and metaphors and discuss theories, popular conceptions, sayings, and images about the topic. Finally, each member should consider what is not written in the memories (but what might be expected to be) and rewrite her own memories given this review and analysis.

The third phase involves reappraisal of the memories and analyses within the context of a broader range of theories. All the memories across different cue words (e.g., happiness, sadness, guilt in Crawford et al.'s work or air, fire, and water in our study) generated by the collective are compared and contrasted. Earlier theories are reevaluated in light of later theorizing, and these revised theories are examined in light of particular theoretical positions and commonsense understandings. New understandings are recursively used to reappraise initial analyses of the memories.

Referring to the injunction to avoid autobiography, the rules facilitate writing a description of an event as opposed to an abstraction or summary of an event, something that is easier to accomplish in the third person. These rules help to avoid interpretation and justification while writing a memory, and the result is a memory that is open to subsequent or concomitant analysis.

Memory-Work and Narrative Inquiry

The written memories produced through memory-work share many characteristics with the stories of personal experience produced through narrative inquiry. However, in narrative inquiry, other lives are being studied by a narrative researcher who is describing, collecting, and telling stories of individual lives (Clandinin & Connelly, 1995). Like stories or narratives, memories involve a sequence of events, a storyteller and an intended audience, and like memories, narratives reflect events and experiences important to the teller. They reveal social, community, and institutional values and realities (Cole, 1991). They help us understand the meaning of everyday life. Narratives play a central role "in the formation of the self and in the construction, transmission, and transformations of cultures" (Witherell & Noddings, 1991, p. 3).

> Stories people tell about themselves, about others and about events or experiences
> seen or heard reveal to the listener those cultural, social and personal values and

behaviors which are salient to the speaker's identification. This is so because people make choices about what is reportable in accordance with their own views of the cultural models and values they hold to be inherent in their own psychological economies. (Brunn, 1994, p. 2)

The telling of stories, like the telling of our memories, as Clandinin and Connelly (1995) note, "is a way, perhaps the most basic way, humans make meaning of their experience. . . . Active construction and telling of a story is educative: the storyteller learns through the act of story telling. . . . The story is reshaped and, so too, is the meaning of the world to which the story refers" (pp. 2–3).

Memory-work is similar to the traditions of narrative inquiry for, like memory-work, narratives reflect

> our inward journey that leads us through time—forward or back, seldom in a straight line, most often spiraling. Each of us moving, changing, with respect to others. As we discover, we remember; remembering, we discover; and most intensely do we experience this when our separate journeys converge. Our living experience at those meeting points is one of the charged dramatic fields of fiction. (Welty, 1984, p. 112)

Memory-work's similarity to narrative inquiry is further illustrated by the position of the researcher. The researched and the researcher are both part of the process, and "the two narratives of participant and researcher become, in part, a shared narrative construction and reconstruction through the inquiry" (Connelly & Clandinin, 1990, p. 5).

Despite the similarities, memory-work retains a distinctive characteristic that sets it apart from the narrative forms of biography and autobiography. Carefully crafted stories about people are full of omissions and evasions. They are edited summaries of lives from the perspective of the present, often narratives of causality and logical sequences of events that resonate with our present views of events and ourselves (Evans, 1999). In contrast, the discrete memories generated through memory-work are intended to illuminate the moments of a life as opposed to causal strings, and as such, these memories are much easier to compare and contrast than tightly woven autobiographical summaries. While some of our experiences with the elements might have been included in autobiographical narratives, the rich detail and complexity of the discrete memories would have been lost. The process whereby we become the person we are may be studied because the complexities of our early experiences survive in these discrete memories. These complexities include the effects of economic class, ethnicity, gender, sexual identity, physical limitations, experiences of inclusion and marginality, and values and ideology.

We note however, that we did not completely avoid autobiography. Once we had written all of our memories and analyzed them across ages and elements, we decided to analyze memories by individual. This gave us a broader context and helped us answer questions about what was missing in a memory, but it shifted the focus to autobiography. We explain this shift in method more fully below in the section entitled The Seduction of Traditional Analysis.

Given this distinction between discrete memories and auto/biography, memories are still "reflections that return the past to the present" (Grumet, 1990, p. 322). To revisit a memory is to "see oneself seeing" (Grumet, 1990, p. 322) but through a new construction. We can never recall a memory as it was because we are no longer the same individuals whose experiences created the memory. Therefore, memories are constructions and reconstructions based upon who we were, are, and are becoming.

A final point to be made about the method of memory-work concerns the limits of language and the need to engage collectively when using the method. Haug's socialist-feminist collective analyzed their past experience with the aim of discerning their own participation in their subordination and oppression as women. They recognized, however, that they were constrained by the language of the patriarchy. As the collective noted, "we found ourselves speaking, thinking, and experiencing ourselves with the perception of men, without ever having discovered what our aims as *human beings* might be" (Haug, 1987, p. 36). Individually, it is extremely difficult to rethink and question ideas and ideologies that shape and become a part of our being; collectively, though, the task becomes easier. Each member, with her differing perception, has the potential to see what was previously unseen and to say what was previously unsaid. New ideas emerge from collective scrutiny of memories and from the continuing struggle to be more conscious of the use of language.

Our Experiences with the Method

After deciding to use Haug's (1987) work as a method and theory to explore our relationship(s) as women to science, a difficulty arose in deciding on what kind of episodes or events to write about. In selecting specific cues for memories, Haug (1987) cautioned that clichéd cues would lead to clichéd memories. For example, Kippax et al. (1990) found that cues such as "first love" produced clichéd and unproblematic memories, whereas "touching" produced more complex and varied memories (in Crawford et al., 1992). After much discussion we decided to focus on our relationships to the natural world and decided that the classical elements (air, earth, water, and fire) were complex enough to avoid cliché and simple enough to stimulate a variety of different early memories. We reasoned that memories cued by the elements would help us examine our socialization in relation to the natural world. Insights regarding our socialization would in turn give us insight into our relationship to science because science is tied to observation of nature.

Data Collection

We used the basic ground rules established by Haug (1987) and Crawford et al. (1992):

1. write one of your earliest memories;
2. of a particular episode, action, or event;
3. related to the identified cue;
4. in the third person;
5. in as much detail as is possible;
6. but without interpretations, explanation, or biography.

Following Crawford et al.'s (1992) strategies, we read the memories aloud to the group, asked questions, clarified details present and missing, and added needed context. Discussions following this protocol were audiotaped and transcribed.

We also looked for clichés, contradictions, and popular conceptions. Haug (1987) suggests rewriting the memories after this process. Crawford et al. (1992) found this unproductive. Not knowing the potential utility of this strategy for our own work, we decided to go along with Crawford et al.'s (1992) critique and did not initially rewrite memories. However, discussion and analysis of particular memories over four years prompted us to consider formally rewriting selected memories. The criteria for choosing memories were that they generated extensive discussion and as a result were elaborated by the rememberer and understood more complexly by the collective. These memories evolved in the life of the collective, and many of the transformations that appeared in the rewrites are evidence of a new perspective.

Preliminary analysis of our earliest memories of air, water, earth, and fire led us to collect more data. We thought that another set of memories might give us more insight. Two members of the collective expressed a strong interest in using "tree" as an organic cue (transcript 5/5/94). We also decided to include two sets of later memories, from adolescence and young adulthood, speculating that we might see changes in our relation to the natural world. We were particularly intrigued by the emerging work on the silencing of girls' voices during their transition from late childhood to early adolescence (AAUW, 1992; Brown & Gilligan, 1992; Gilligan, 1982; Taylor, Gilligan, & Sullivan, 1995; Sadker & Sadker, 1994).

The format of our memories differed among members of our group and sometimes strayed from Crawford et al.'s (1992) writing rules. Our memories varied in length; they varied in the amount of detail we included, and some of us generated a number of memories in response to a particular cue. Some of us used underlining or capitalization for emphasis in our memories, and some included dialogue. A few memories were written in the first and not in the third person. In addition, the link between the memory and the cue varied from the obvious and the direct to more obscure associations.

Strategies of memory generation were diverse within our group. Most of us believed that we had trouble remembering and frequently did not write our memories until just prior to meetings. One commented that she thought about the cue, and when she finally sat down, she just came up with a memory. Another used meditation as well as sensory imagery (e.g., imagining the smell of fire) to help her recall memories (4/11/94). For a third, "the pressure of coming in the door" to the session enabled her to recall specific details about her memories (5/5/94, p. 6). The cues for generating memories were expanded beyond the literal and obvious. For example, one person saw earth as dirt in a house and another, as sand in water.

At the beginning of our study we found that each of us in one way or another edited or censored our memories to make them conform to some unconscious or unspoken notion of group expectation. Anna confirms that:

> I think our stories are getting more and more alike in form. And, I think it has been more conscious on my part. I recognized when I started writing this memory that I used adjectives and adverbs in ways that would sound like the writing of everyone else. . . . There might be a phase then where you . . . are feeling more pressure to conform or to make your stories ones that we can now analyze. (5/5/94, p. 9)

This pressure to conform was evident in our selection of memories. Bell observed that she selected memories that were focused more directly on science and nature. Also, if more than one memory could be recalled in response to a cue, the more conventional memory might be selected for presentation. Censoring occurred if we had difficulty recalling memories. Bell noted that with a particular memory, "I couldn't place it exactly, and I didn't know how old I was, so I just let it go" (5/5/94, p. 1). Anna indicated that she chose a memory, not because it was the first she recalled, but because she could trust that it was actually her experience.

> We were toddlers eating dirt. I know of the story, but I don't know if it is my memory or if somebody told me the story. The things I remembered had to do with my mother wiping my face, brushing my diapers and the warmth of the siding of the house and the cement. I don't know who was there. I could taste the dirt and I could remember tasting dirt. (5/5/94, p. 5)

Rae raised another aspect of censorship in the following excerpt from a transcript.

> My feeling is that, for me, it is having somehow gotten it in my head that we are going to do it this way so I do it that way. If it says third person, if it says cut out the biography, that's how I do it. But if it says don't make judgments, don't analyze, then I try not to even though I've made judgments, like the hot air ascension with my brother. I can't think about that memory without knowing perfectly well that I was

taking care of him. My brother tried to do something and it didn't work, and I felt terrible for him. Now, in part, that is a feminist issue. (5/5/94, p. 9)

The hot air ascension Rae refers to follows:

Rae, Child, Fire~Paper Balloon Memory: *Rae is about 9. She stands in the empty lot next to their house. It is fall, the leaves dry and brown, rustling underfoot when she and her brother Fred walk to this spot. He lights a tiny fire of leaves under a balloon he has made by folding a newspaper into a ball with an opening in the bottom. He explains that the air will heat and raise the balloon into the air. It does not happen. The paper balloon does not rise and Rae feels sorry for him.*

One of the rules we adopted for writing memories was to refrain from judgment or analysis. Of course, *"feels sorry for him"* is a summary judgment of her feelings, but Rae does not go further and equate feeling sorry with "taking care." She censored her analysis, even though she knew it defined the memory for her. In this case though, censorship leaves the memory more open to collective discussion and interpretation. Including analysis and judgment can undermine collective discussion because judgments can lead us to discuss the memories in clichéd terms rather than move outside established meanings and constructions. Our acknowledgment of all of these aspects of censorship early on in the process helped us recognize our self-imposed prohibitions regarding the form and content of our memories. As we developed our use of memory-work, we were able to confront these problems and move past them.

The question of how we accessed our memories emerged in these discussions of censorship. Bell and Rae used a strategy for access we termed "cognitive recall." For Bell it was a sequential and logical process of linking the element to the age in order to access a memory. For example, she would think about where she was at a particular age and then think about the element within that age context. When Rae searched for an early air memory, she thought about her knowledge that hot air rises. That brought her brother's attempted paper balloon ascension to mind.

When Anna described her memory of eating dirt as a toddler, she noted that she could actually taste the dirt when she accessed the memory. Anna's sensory access was intriguing. As an experiment, she

listed the memories that were coming to mind and what triggered the memory. We were talking about imagery. For this element, earth, my memories were 80% sensual. When trying to access a memory, I closed my eyes and thought about earth. I heard horses or deer running on a path. All those things came through. When I thought about smelling earth, tons of memories came to mind. Many were garden memories, because I had my garden and there was always a garden at home. And then touch, many memories came from touch. (5/5/94, p. 6)

Anna's experience prompted us to broaden our definition of imagery. For Anna and Sue imagery was visual. Their images also included other senses. About one early earth memory, Anna noted that "I forced myself to forget the image. I had nothing to substantiate it and I didn't trust myself to remember anything from when I was a toddler" (5/5/94, p. 7). Sue's image memory was based on a picture in her parents' photo album and not, she thought, on the actual experience. She did not "trust the memory. This other memory I know happened. I remember it very, very vividly" (5/5/94, p. 7).

Throughout the process of generating memories and asking questions about them, the metaphor of sediment kept recurring. Haug (1987) and Crawford et al. (1992) employ this metaphor a great deal in their discussions. Sediment is what falls to the bottom of a liquid; it is typically found in layers; it documents time and yet disguises what it covers. Applied to memory-work, sediments are the evidence of social construction and our participation in that construction. Sediments reveal established meanings and traditional roles, but they also have the potential to lead us to what is forgotten.

Early in our work, we recognized that we were generating at least two different types of memories: events and amalgams. The event memories, like episodic memories (Tulving, 1972, 1983), are linked to a specific point in time and a specific event. The amalgams are somewhere between Tulving's (1972, 1983) episodic and semantic memories. Semantic memories metaphorically hold general knowledge that is built up from repeated experiences. For example, an individual's knowledge of how to use a hammer comes from repeated experiences with a hammer. However, those experiences are implicit and not consciously recalled when the hammer is used. Our amalgams represent repeated experiences, and though each specific experience is not recalled, we know the memory represents a collection of experiences or an amalgam of repeated events in our lives. Bell's memory of adolescence, generated in response to the cue "earth," is an example of an amalgam.

Bell, Adolescent, Earth~Growing Rocks Memory: *Her mother always required that she and her brother help dig up the garden every spring. She hated doing it and since it was her mother's hobby anyway—why did she have to do the dirty work before her mother started the planting? She had to use a pitchfork first and dig way down to loosen up the soil and make sure to shake as much dirt as she could from the grass clumps. The thing that was always amazing and what they joked about every year were the rocks. Every year it seemed like she dug up a ton of rocks, and there wouldn't be any boulder-like rocks left in the garden. The soil was smooth and clean, and you could drive the pitchfork in and not hear that clunk or feel that bone-shaking vibration when you hit a boulder. Sure enough though, year after year, those boulders grew back, and every year there were just as many of them and they*

*were just as big and some took one half hour to just find the edges of the boulder so
you could sense how large it was and whether it was even possible to pull it out of
the ground.*

This memory represents all or at least many of the times that Bell and her
brother dug up the garden. The details are strung together from separate events and
emerge into one amalgam. We wondered why, besides sheer repetition of a recurring
and frequent experience, we accessed and constructed amalgams? While discussing
an early earth memory, Anna said:

> I can take an amalgam and create necessary details around it. It must be true be-
> cause I know from family history this is how it was. I know how big I had to be in
> order to physically accomplish the task. I know that Grandma and Grandpa were
> there because we didn't go there much after my aunt and uncle died. All these
> things that you create in an amalgam are things most likely to occur. They are not
> actual. This is how I imagined this. If I don't have a specific episodic memory, I
> create the amalgam. (5/5/94, p. 12)

In an earlier discussion, Anna observed that the amalgam may provide a narrative context around a specific memory. In this manner, amalgams are sediments or
social constructions assembled from many memories. We might have a memory of,
for example, digging up the garden, but because Bell did this a number of times,
there is no specific event that is accessed. The only way then to communicate or
articulate the memory is to imagine what must have happened, that is, to glean details (or sediments) from all the "digging garden" experiences. Rae noted, "If we
want details, by golly, a lot of them are going to have to come from lots of similar
events in the past" (3/3/94, p. 16). Amalgams, then, are sediments linked to multiple memories.

The Seduction of Traditional Analysis

As we began to think about the entire corpus of memory narratives, it became
abundantly clear that data and analysis are intricately interwoven. In fact, the process of analysis has been and is continuous. Each time we return to the memories,
we gain new understandings and insights, and these insights change the way we
think about the memories. This is what Haug (1987) describes as the recursive nature of memory-work.

Coming to terms with the idea of analysis as a continuous and recursive process
was a first step in becoming conscious of the conflict between our traditional training and a method that contradicted essential tenets of that training. The next step
was coming to terms with our impulse to objectify the data; our impulse to separate
the researchers from the researched. When we began to consider the memories as a

whole, we had only begun to scratch the surface of the memories. In Haug's terms we were largely oblivious to the social structures that shape our lives. We saw commonalities in our memories; we saw patterns, but in our initial attempt at describing our findings, we had only recapitulated our socialization.

For example, in some of our memories we were gregarious, outgoing girls, curious about the natural world, but as we grew into adolescence, peers took center stage and we spent less time exploring the natural world. This set of observations supported our theoretical knowledge that this was the expected movement from childhood into adolescence. However, within the context of memory-work we had not yet begun to question the expected or "natural" in our memories; we had not yet begun to see that what we thought was natural was socially structured and that we subordinated ourselves to these structures. Just as Haug (1987) describes the initial attempts at analysis in her collective, it seemed to us as well that "everything was relevant, possible, random, impervious to analysis" (p. 59). Haug's collective also recognized that their "initial discussions of the stories yielded little more than a repetition of the interpretations demanded by our different positions within the social hierarchy" (p. 59). Their strategy "was to make necessity the mother of invention, in other words to view the initial discussion process as a peeling away of the layers of material sedimented in our minds" (p. 59). We adopted this strategy as well but only after a frustrating process of subjecting the stories to traditional modes of analysis. In retrospect, this process forced us to confront the influence of our traditional training.

Our traditional analysis began with a table of more than 90 memories (see Appendix A) organized by time (child, adolescent, adult) and person. The table of memories resulted in ordered data, but we were still searching for answers to our basic question: What do these memories tell us about the natural world, ourselves, and girls in science? Two salient dimensions of the memories emerged; the presence of others in the memory and the degree to which the memory was focused on the element. We wondered whether focus on an element such as air diminished with the number of individuals present in a memory. Because both dimensions were continuous, we decided to graph them in a two-dimensional space. We agreed on a notation system for the memories, but we could not reach consensus on graphs that each of us produced. They were extremely divergent depending on how each of us had defined the dimensions. We tried constructing one graph as a group and still could not reach consensus. We reluctantly admitted that our attempt at categorizing, naming, sorting, and quantifying the memories had failed. We went back to the proverbial drawing board and asked ourselves: "What are we studying?" We wondered if we could really examine our relationships to science through the cues air, earth, fire, water, and trees. We became conscious of our seduction into traditional analysis, and we returned to the memories with a fresh gaze.

Memories and Individuals across Time

We looked at the memories with a developmental perspective, but some contradictions emerged from this analysis, and they led us to reflect on the "manufacture" of development as opposed to a "natural" course of development. Taking a developmental perspective modified the method described and defined by Haug (1987) and Crawford et al. (1992) and was contradictory to it. Memory-work prescribes and builds upon snapshots of a time and place, and analysis emerges from this. We, however, were concerned with a perspective on change over time. That is, how women's relationship to nature and science develops. The demands of our question led us to extend the method, but this extension meant that we might contradict the method. By looking at change over time, were we not considering the possibility that our relationship to nature and science could be described through a linear and causal perspective? We decided that we could not ignore the passage of time in our relationship to nature. However, we carefully scrutinized conclusions that hinted at any causal links between, for example, a childhood memory and a young adult memory. Interestingly, we found that as we became older, some of us moved further away from our connections to nature. We observed that our intimate connections to nature as young children bore little resemblance to our relationship to nature as adults.

We also modified the method by examining each individual's memories as a whole. In contrast to Haug (1987) and Crawford et al. (1992), we reread all of an individual's memories from early childhood through adulthood. As we moved through the individual narratives, contextual issues moved to the fore. Rural and urban contexts, class membership, sibling and parent relationships, sexual identity, spirituality, and ethnicity all emerged as important factors in our memories. We felt that this broader qualitative analysis much more accurately captured the "complex pattern of an individual life without violating the integrity of the life or dehumanizing the individual" (Kotre, 1984, p. 3).

From this broader analysis, themes of individual lives emerged more clearly than before. At the same time, some of us felt vulnerable when describing the contexts in which memories were embedded; others admitted censoring what we told the group. We all identified issues that were particular to our lives, such as need for control and chronic illness. At the same time, we consciously avoided telling simple causal stories of our lives. As we gained more knowledge of each other's lives, we asked different questions, more incisive questions, and we challenged the stories that we told to one another.

We became more sophisticated in our analyses as we gained experience with the method. We uncovered many of the "missing" or unwritten elements of our memories. Our analysis was collective and oral, and it enabled us to see more clearly our layers of socialization. In many ways, revisiting the memories conforms to Haug's suggestion that all memories be rewritten following collective discussion. At the same time, however, examining each individual's set of memories is a contradiction of Haug's injunction to avoid autobiography because the focus of memory-work is on the social and not on the individual. However, this individual focus stimulated questions that were not apparent to us until we saw the weave of our individual lives and knew when and how to ask about issues we largely keep to ourselves. Exploring each woman's entire set of written memories cast new light and facilitated our move from one understanding of a particular memory to a broader more complex understanding.

The benefit of analyzing one individual's set of memories is illustrated in Rae's red boots memory of standing in a street gutter on a rainy day and watching water rush over the front of her boots. She described her interest in the water's flow and its effect on floating sticks. The initial reaction of the group was that this memory was a systematic exploration of water. However, when we began to question our own expectations about what we would find in our memories (e.g., "science-like" activities), the group's understanding of that specific memory changed as did Rae's account of it. A missing element of the memory was Rae's attachment to the red rubber boots she was wearing in the memory. In her retelling of the memory, Rae remembered the joy of feeling water rushing over the rubber boots, the feeling of water flowing over feet standing on concrete. She emphasized that she was not systematically exploring water, rather she was rapt in the sensuous enjoyment of water.

This "seeing" of the sensuous marks a shift where we began to move beyond the obstacle of our then-current ways of understanding. Rae's initial telling of this memory and our collective analysis of it are products of the dominant norms and values shaping our perceptions. We imposed a traditional view of science on this memory and a traditional view of children engaged in science-like activities. However, once we began to disentangle the memory, to question our judgment of it, we saw a new way of relating to the natural world, one that involves sensuous embodiment rather than exploration and manipulation. Haug (1987) expresses the shift well:

> Once we have begun to rediscover a given situation—its smells, sounds, emotions, thoughts, attitudes—the situation itself draws us back into the past, freeing us for a time from notions of our present superiority over our past selves; it allows us to become once again the child—a stranger—whom [sic] we once were. With some astonishment, we find ourselves discerning linkages never perceived before: forgotten traces, abandoned intentions, lost desires and so on. By spotlighting one situation alone, we learn to recall and to reassess history. (p. 47)

Emergent Analysis

As our analysis progressed to examining individuals' memories over time and analyzing our research meeting transcripts, consistent findings began to emerge. The concept of personal science emerged from our consideration of the sensuous, and evidence of sensuous experience in the natural world came up repeatedly. This repetition flagged its significance. Creativity, use of metaphor, family, and power were themes that emerged again and again in our analysis of the memories. Discovery of these themes led us to reexamine the memories, our earlier analysis of them, and the transcripts of our meetings. We coded the memories repeatedly and independently for these thematic categories. We also coded the transcripts of our meetings, and this reinforced and modified our tentative themes. We collectively discussed the fit between the coding and the generalized themes. We have examined and reexamined the relationship between evidence and themes in our discussions and have situated our findings in the context of related research, which, in turn, provides new ways to examine our data and findings. The process of analysis and search for related ideas is never-ending, and our discussions have continued into the writing of research papers and this book.

Memory as a Construct

Memory-work carries with it a number of assumptions about memory that need to be clarified before we move on to a discussion of our findings in the next chapters. We describe traditional notions of memory first and then discuss the position we take on memory throughout the remainder of this book.

Memory, as a psychological construct and neuropsychological construct, can be described in a myriad of ways. The computer metaphor, still popular as a model for driving research, has generated terms such as storage, retrieval, and capacity (see Atkinson & Shiffrin, 1968; Norman, 1970). Long-term, short-term, and sensory storage are terms used to describe location and the length of time memories are stored or used to perform particular cognitive operations (see Bjork & Bjork, 1996, for an overview of work in these areas). Within this literature, there are descriptions of how we access memories from long-term storage and how we work with memory in short-term storage. There are also descriptions of how we organize the information that is theoretically stored in our memories. As briefly noted earlier in the chapter, memories can be described as episodic, linked to a particular moment in time, or they can be described as semantic (Tulving, 1972, 1983). A semantic memory represents the general knowledge that accrues from the various episodes or particular experiences that make up, say, our knowledge of how to read. Memory can also be described as procedural, which is the knowledge of how to do something such as ride a bike or rebuild a carburetor (Anderson, 1983).

At a psychoneurological level, we have begun to locate those places in the brain that appear to store particular kinds of memories and areas that are connected in terms of brain function. Damage to the hippocampus and some connected areas impacts episodic memory but not the ability to learn new skills, while basal ganglia injury affects skill learning and leaves episodic memory intact (Engel, 1999). Other research links the formation of neural connections with incoming sensations and as sensations and experiences are repeated, the connections are stamped in, so to speak (Carter, 1998).

Memory, even within the traditional research literature, is regarded as a theoretical construct, and despite the myriad ways of describing and defining it, memory does not exist in a material sense. However, few researchers, we suspect, would be willing to entirely dismiss the idea of memory, and, in this book, we include ourselves among those not willing to give up on the idea of memory. Memory has reality in the sense that a chair has reality when you bang your knee on it. It seems to be there, and so it is with memories—they seem to be there and we seem to have them. So we admit to the reality of memories, but there is an assumption about memory that we bring into question, namely, the idea that the past or, more particularly, our memories of the past determine who we are in the present. For example, it is quite common to hear an individual say that "event X caused me to behave in this manner," or that "who I am now is based on the events of my past." We critique this determinism and argue that our memories represent who we are at the moment in time that we recount them. The memories that we recall reveal, in part, who we are now, the values that we hold, and the beliefs that we have about our lives. In short, our memories are a record of our socialization, but they do not trace the inevitability of our present selves.

While much of the traditional research on memory has practical application in particular contexts, it does not, nor is it intended to tell us much about the connections between our memories and who we are as cultural beings. Additionally, the research does not lead us to consider the poststructural perspective, namely, that our conceptions of memory are produced from the very discourse used to describe memory. Our culture produces a very particular way of looking at the world, and this in turn produces particular methodologies. These methodologies both create and give substance to objects (such as memory) and create ways of looking at the particular objects we study. However, as Freeman (1993) reminds us, we do have experiences that correspond to the discourse of the culture. Therefore, we cannot easily dismiss these experiences or we descend into nihilism where everything is nothing. For example, a person might remember what she had for lunch yesterday but not what she had for lunch two days ago. She is experiencing the effects of short-term memory, and even though she may realize that short-term memory is a concept produced by a particular discourse, that does not negate her actual experience. Thus, even though we may strive to live consciously in the culture we create, we often have to make a leap of faith and believe that while something may not have substance in a material sense, it still has an impact, and, in the case of memory, may still be a worthwhile concept to study.

Our experience with memory-work showed us that when we remember or narrate a story of a particular event, the story of that event is only one of many stories that could be told about that particular moment in time. For example, when one of us narrated a story of driving with her father to Boston on a hot summer day, there were many, perhaps an infinite number of, stories that could be told about that ride. She told a story about the play of her arm in the wind hanging out a car window because in the present she was searching for a memory about air. There is nothing inevitable or determinant in that memory though it does contain evidence of the present. Another aspect of the memory is that it would not have been included in an autobiographical rendering of the narrator's life, and that is true of most if not all of the memories recounted in this book. Relying on the method of memory-work, we intentionally tried to avoid telling determinant and linear autobiographical narratives of our lives. It was our intent to interrupt and subvert the assumption that the past determines who we are now. However, given this, can we assume that the discrete memories recounted in this book are somehow more accurate depictions of our past? These memories certainly reflect how we thought, sensed, and felt the world around us as children, adolescents, and adults, but they are still subject to interpretation. We can never accurately record the moments of our lives because we are always constrained by particular ways of looking. Memory-work can only provide another way of thinking about who we are and how we have been socialized.

The discrete memories retrieved through memory-work were determined by the questions we were asking, the context of listening to each other's memories, and the selves that we believed and believe ourselves to be. Thus, the memories reflect who we are now, but they escape the justificatory tone that we use when we tell the stories of our lives. When we tell the stories of our lives, they are stories that end where we are now. The events that we choose to narrate and the ways in which we narrate them lead neatly to the selves we are now. We habitually and perhaps unconsciously omit those events that have little to do with or perhaps contradict the selves that we portray in our narratives. These things are forgotten; they do not work well in the narrative.

However, it is foolish to take a definitive stance and argue that the past does not exert any influence on the stories we tell. Experiences of sexual and physical abuse, in short any traumatic or compelling experience, can influence a life and therefore influence the stories we tell. As Freeman (1993) wisely notes, "we cannot think of this state of affairs in either-or terms but must instead embrace what I called a 'both-and' perspective, in which we are willing to read the text of a life both backward and forward" (p. 226).

We conclude this chapter with a quote from Adrienne Rich. We think it captures the essence of our project and the significance of memory-work to our lives as women in science:

As the hitherto "invisible" and marginal agent in culture, whose native culture has been effectively denied, women need a reorganization of knowledge, or perspectives and analytical tools that can help us know our foremothers, evaluate our present historical, political, and personal situation, and take ourselves seriously as agents in the creation of a more balanced culture. (Rich, 1979, p. 141)

PART TWO

CHAPTER 4

Making Sense

Linguistic meaning is not some ideal and bodiless essence that we arbitrarily assign to a physical sound or word and then toss out into the "external" world. Rather, meaning sprouts in the very depths of the sensory world, in the heat of meeting, encounter, participation.

—Abram, 1996, p. 75

The language of understanding makes obvious the centrality of the senses in knowing. When someone makes a judgment about the quality of a story, the truth of a reality, the question is "Does it make sense?" The direct meaning is "Are the senses awakened?" A story that makes sense by definition requires the participation of the body. When something makes sense, we see the point; we have touched the essence; I say that I hear you; we say "It feels right."

Making sense is necessary to science, although in the past century much of science has become abstract, separated from personal knowledge of the physical. Nevertheless, understanding and interpreting sensory information is an important part of science, whether the natural or social sciences. This kind of information is evident in many of the memories that seem to us to hold the keys to understanding the development of our relationship to nature and perhaps to science. The senses, our bodies, provide access to the natural world, and our embodiment in it is, of course, shaped by the social interactions that are also part of nature. In this work we do not use embodiment to mean personification. Rather, we refer to embodiment in terms of the body as a site for making meaning. In this chapter, we are particularly focused on sensation, which begins in the body.

Although conventional wisdom speaks of cognition as if it were an independent enterprise of mind, much of our understanding is rooted in the body. Johnson (1987), for example, maintains that much of our elementary comprehension is based on our knowledge of the structure and orientation of our bodies. Our reckonings of "inside" and "outside" may be based on our experience of our bodies as containers into which we put food or fingers. These bodies are spatially bounded and may serve as models for "in" and "out," which we use intuitively in constructing categories. Perhaps our notion of processing sensation in "higher" centers begins in our experience of our sense organs, such as eyes, ears, nose, and tongue, being concentrated in the head. And the head is at the top of the body, "higher," during most kinds of activity . . . sitting, walking, standing, for example. Similarly, Johnson points out that our knowledge of "balance" as a concept begins with our bodily experience of balance and its necessity to successful physical activity. We learn how to balance our heads on our bodies. We learn about balance also in terms of physiological limits; we know when our bodies are too hot or too cold; we know if our mouths are too dry or we are thirsty. We know if there is "too much" or "not enough," and we learn the concept of balance in offsetting these conditions. At the same time exploring balance leads us to test limits, reframing possibilities.

The study of infants offers insight into the complexity of the relationship between sensory-based perception and social interaction. Arguably, infants are social beings from beginning. One aspect of their sociability is highlighted by the observation that six-week-old infants appear to differentiate between their mothers and familiar objects (Trevarthen, 1979) and that slightly older infants respond differently to seeing dolls and seeing active females, to hearing sounding objects and speaking people (Legerstee, 1991). Certainly infants respond differently to sensory input without social stimuli and sensory input accompanied by or embedded in social interaction. Embodiment, in the sense of the body as a site for making meaning, necessarily entails the social interactions that inform perceptual experience.

In our analysis, memories were coded by sense. The sensory content of each other's memories contributed to our judgments about the salience of various parts of the memories. For example, the image of people buried alive, the sound of a tornado, and the feeling of an earthquake were striking. For the collective, some sensory modalities have greater impact than others in the memory narratives recounted in this chapter.

Here we discuss how the sensuous in our memories provides the observations, the data, for understanding how the senses meld with the social in influencing our connections to nature. We are particularly focused on the ties between sensuous experience in its social context and the emergence of a personal science, which is embodied during everyday activity. Finally, we suggest that ordinary embodied experience, neither necessarily systematic nor abstract, may hold keys to girls' development in relation to nature and perhaps to their interest (or lack of interest) in science.

Personal Science

Our cultural grounding in science as positivism initially prevented us from recognizing the personal science in our remembered experiences. We tend to associate "real" science with distance, symbolized by the white lab coat and the closed laboratory door, and with formal hypotheses written down and compared against observations from experiments, whose results are also written. Hubbard (1988) reminds us that other views of science are possible. Typically, natural scientists look at small pieces of nature in pursuit of objectivity. The isolation of the part is thought necessary to understanding its function unencumbered by its interactions in its usual setting, a practice Hubbard calls "context stripping." In contrast, she calls to our attention Freire's (1985) notion that our understanding of reality depends on "the indispensable unity between subjectivity and objectivity in the act of knowing" (p. 51). The primacy of objectivity in traditional science makes the importance of the subjective as a part of context difficult to recognize.

Hubbard (1988) pursues the question of why some approaches and contexts are regarded as important to science and others are not. She asks specifically "why certain ways of systematically interacting with nature and of using the knowledge so gained are acknowledged as science whereas others are not" (p. 14). She speaks here of the curious distinction between elite science and the science of women's gardens, kitchens, and nurseries, where knowledge is gained, passed on, and remains devalued and unrecognized as science. Only recently, for example, has the history of women's domestic medicine been recognized (e.g., Allured, 1992).

Knowing gained and applied in domestic settings typically is not considered science (Barr & Birke, 1998). As mentioned in chapter 1, "Introduction," our collective named the process of acquiring that knowledge "personal science," and we believe it begins in childhood though it continues through life as the settings in Hubbard's list imply. We begin with a memory that exemplifies adult personal science, which can be identified readily as learning in a domestic setting (the front yard) and which is clearly embodied as the references to taste demonstrate. In the following memory, a woman's knowledge of the almond tree is anchored in memory by the sensation of taste. Here the flavor of green almond is metaphoric and central to learning through new experience in a new place. The seasons of the almond tree and the ripening and aging of its fruits from astringent to mellow provide a taste of the experience.

Anna, Adult, Tree~Almonds Memory: *Who would have guessed that those little green balls that kept falling from this tree had almonds inside of them? The fall of 1972 brought a new job teaching fourth grade and the newness of owning a home and the beauty of the almond tree. The almond tree took up the entire minute front*

yard of the house on 1725 Juniper Avenue in Silver City, New Mexico. Anna har-
vested the meager crop (they say the trees froze last February after a long warm spell
that prematurely budded the tree), tasted the green almonds, and waited for another
year to learn more about this tree. Years did educate Anna about the worth of the
spring buds, bright green early March leaves, abundant crops of almonds, neighbor-
hood chats about and under the tree, and the value of aging the almond before con-
suming it.

We found that Anna built on her childhood experience of planting gardens and observing their changes over time. In New Mexico, that earlier knowledge was extended to a new horticultural variety. This scientific learning occurred in the personal sphere of home, season by season. Furthermore, the social entailments of the setting—learning from neighbors under the tree, the ties to a new job where children are learning from her, the responsibility of owning a home—are clearly an important part of the context for Anna.

Traditionally, learning about plants, their requirements, and changes over time in the family garden is of no scientific consequence; it does not take place in a laboratory; conditions may not be controlled; emotional connections to the learning experience are not necessarily stifled. However, we all acquire a personal science. We observe, though we usually do not record our observations; we see the results of changing conditions; and we generalize over our observations during years of experience. This is part of what science entails: acquiring knowledge of nature over time. Personal science, like Anna's, can take place in the domestic sphere and continue to expand over decades of experience that include personal feelings of aversion or enthusiasm for the phenomena, pleasure or distaste for the sensuous aspects of contact, and warm or cold regard for the people who are part of the experience.

In each of the sections that follow, we will return to personal science in discussing the evidence from sensory details of the memories. The sensory is important in our analysis in several ways. Particular parts of the memory become salient because, first, their sensuous impact on members of the memory-work collective as individuals and as a group is strong and, second, vivid sensuous connections suggest the importance of the body in relationship to nature and personal science. The more we focused on the body as a conduit for knowing, the more we saw our physical connections in nature. Thus, we turn now to a range of memories in order to discuss integration of the senses in some and demonstrate the preeminence of the individual senses in others.

Integration of the Senses

For some memories, the sensory complexities of the whole experience provide salience for us. Sometimes we identified with each other's memories when a single sense was described; however, we connected more strongly with the social meaning of the experience as more sense avenues were opened to us as listeners and readers. The integration of the senses invites wonder, joy, amazement, or perhaps horror, surprise, or revulsion in response to the total experience. Perceptions overlap and interact, one sense informing another to draw us into the memory.

Perhaps we can understand the importance of multiple sensory avenues in creating a strongly absorbing account of memory by thinking about an analogy provided by Abram (1996). He offers the work of the magician, who provides meaning for the observers from several cues. More precisely, the magician suggests a seamless reality, bridging the gap for the observer from small obvious parts of the trick, the pieces, to a perception of the complete experience. The pieces are chosen carefully for their potency of suggestion and inference. Similarly, recounted memories employing a range of evocative details grab the listener. We refer to memories that employ that breadth of access as integrated.

Cele, Adult, Water~Glacier Memory: *It was cool on that September day in Alaska. Cele had been persuaded by the locals in Juneau to take a drive around the environs. She had flown over the glaciers when she flew into Juneau, but she was not prepared to come face to face with the living, moving mountain of ice. It was a huge talking, living thing. The cracks and crevices were uneven and rough; it was slowly giving up its hold on the ground beneath. A living replica of the Ice Age slowly giving way to human intervention.*

This memory refers to a long-term natural event. The evolution of the glacier over thousands of years is evoked in terms that suggest an old beast flexing its claws. The human dimension of this phenomenon is clear: The glacier looms personally, face to face. The memory of this encounter of woman with monumental ice is compelling to the reader, in part, because the element, the ice, is portrayed as having a social dimension. In addition, the interweaving of "*ice*" and "*talking*" takes us by surprise; it is an echo of the "*living, moving mountain*" that startled the rememberer. Cele certainly experienced a new kind of relationship in this encounter and came away with a different understanding of the power of the glacier, no matter how slowly it moves. Although a glacier may be simply a large mass of ice to an observer, Cele has been struck by the notion of its animate motion. She speaks of it as "*living, moving . . . talking . . . giving up its hold . . . giving way.*" The movement of a river of ice has become a part of her personal science because she has not stood back, distancing herself from it, as would be the case for traditional science. In the process of analyzing this memory, the response of the group to Cele's close encounter became visceral, that is, felt strongly in the body.

We experienced a visceral response to few other memories. One such is Anna's sauna memory in which many senses are awakened by the description she provides.

Anna, Child, Water-Sauna Memory: *Anna was a preschooler because of the setting: in the basement of the Spring Lake home when Grandma still lived there. The memory is a vivid recall of a sauna, which is usually conducted in extended family units on Saturday. Somehow, it doesn't seem like a Saturday, but details are gone. Every sense is involved as the water sizzles into steam on the hot rocks and envelops naked skin. Details are hazy: Grandma has long, saggy breasts; the pails were left in the sauna (getting hot); children used only cold water; the water was piped up from the lake and smelled like swimming time. The Spring Lake sauna was a very hot one, and I never got to go to the top shelf where the steam was the hottest.*

The visceral response can be evoked by different parts of this memory for different readers. We were drawn into the memory in diverse ways. The smell of the water, the image of Grandma, the temperature in the room, the sound of the sizzle, all represented different points of access and have different emotional weights and meaning. For one of us, a fondness for water drew her directly to the smell of the lake. She wished she were a part of this sauna. For another, internalized and unspoken cultural prohibitions against nudity defined this memory. The image of Grandma's "*long saggy breasts*" provoked a negative reaction. The ensuing discussion of sexism, misogyny, and ageism brought the issue of visceral response into clear focus. We understood how visceral responses to this memory were culturally produced and the way in which memories reveal the socialization process.

In the two memories that follow, many sensory avenues converge. Thus, someone who has not had these experiences can readily imagine them given the opportunity to become conscious within the framework of analysis. The degree of visceral response in the reader/listener depends on her awareness of an emotional attachment to the elements and the integration of the senses accessed.

Sue, Adolescent, Air-Ferris Wheel Memory: *This is early adolescence. The fair is truly remarkable. Lights, music, sawdust under feet, the smell of hot dogs and fresh bread. I am with my parents and we are walking from booth to booth, display to display. I am so attracted to the rides, their lights, speed, and variation. The Ferris wheel is the most intriguing for me, though. We stand in line for a long time and finally our time to ride comes. We sit in the swinging seat; the bar is placed in front of us for safety and I hold on for dear life. We move a few feet off the ground and others are loaded on below. At each stop, the seat swings and I grab the bar tighter. We are finally at the very top. My heart is pounding and I am paralyzed. Any movement by my parents frightens me. They don't intentionally tease me with the movement, but I am truly scared. When all the seats are finally full, the wheel moves quickly though the air and the wind blows both forward and backward as we travel in circle after circle. I begin to relax; I begin to enjoy, I begin to lessen my tight grip on the bar. My parents chatter about all that they can see, and after a couple of minutes I am finally able to comment as well. All that can be seen, all that can be heard, and the movement through the air. Then, we stop just as I have finally become accustomed to this seat, the height, and the movement. Too bad this couldn't last longer so that I could enjoy it more.*

And another memory from Rae:

Rae, Adolescent, Air~Clarinet Memory: *A repeated experience: Rae is maybe 16. She likes playing the clarinet a lot. She practices and practices, sitting on the edge of her bed close to the nightstand with the small wooden lamp her father made and an arm's length from the folding music stand, which is almost never folded up. On its shelf, which dips slightly toward each end, is a thin book of exercises with an orange semi-gloss cover that is stained by the oil from her hands and her teacher's at the free corners and the middle of the binding. The slightly yellow pages inside contain many exercises she cannot play and some that she can. She plays them over and over and is entirely satisfied by her eventual control over her hands, her breath, and the column of air moving through the clarinet. When she plays well, a deliberate evenness in her fingers on the keys, a bright wooden resonance, and the barely detectable thunk of the leather pads of the keys seating over the holes please her no end. Finally, she caps the mouthpiece, takes her clarinet apart, runs the weighted chamois cloth through its wet bore, and smears a little cork grease with its distinctive aroma over the corked joints. Each piece, with its personal familiar grain, rests in its worn velvet depression until the next day. She will be able to recall in her hands the feeling of music played well decades after she stops playing the clarinet.*

These memories contain active observation that is a part of personal science and both illustrate the importance of physical response in our experience. Sue speaks to the attraction of the speed and light of the rides as well as her fear of the height and motion of the Ferris wheel: Her heart pounds; she hangs on for dear life until she becomes accustomed to this strange movement. Rae writes about an everyday experience in physical terms: The importance of control over breath and hands, the sound of playing music well, and the memory of it in her hands. The social context is less obvious in Rae's memory than in Sue's. Nevertheless, she is learning the clarinet with a teacher within a long tradition of performing music that has been written down and shaped into exercises appropriate to the learner. It should be added that within that context Rae is expanding her personal science with respect to the behavior of air. She learns about air pressure as she controls volume, about the condensation of moisture from her breath in the cooling column of the clarinet itself, and she learns about the effects of interrupting the column of air to produce pitch, and other properties of sound. This is not the distanced stance of traditional science, setting up experiments with closely controlled conditions in the physics laboratory of school.

Vision

For some it is easier to see the importance of the physical body in making meaning by examining memories in which salience is attached to a single sensory avenue rather than a whole suite of senses. We begin with vision because among hominids it is probably the preeminent sense providing information about the world and how to respond. Sacks (1995) in his account of a man, blind at an early age, who gains his sight as an adult vividly portrays the complexity of the lighted world. The importance of perceiving depth, color, movement, and much more is clear to those who acquire sight only after decades of life without it. The visual field provides a rich variety of information that must be processed in more ways than we are aware, whether we are walking down the street, understanding the meaning of light and shadow in the motion of leaves, or making everyday decisions about human life. Similarly, in memories in which visual cues are important, a limited amount of description elicits in the reader an imagined visual field that extends complexly beyond the words.

> Sue, Adolescent, Water~Bikini Memory: *It was a beautiful summer day. The swimming pool was full of people, and my friends were gathering for a day in the sun. I was probably 15 and had looked forward to wearing the new swimsuit. It was spotted like a leopard and two-piece. As I was only slightly endowed up top, the plastic bra inserts that came with the top really enhanced my perception of my beauty.*
>
> *My typical pool entry strategy was to put my toes into the water first to test the temperature. Then I would gather all my strength and willpower, hold my breath, and jump feet first into the deep end. I entered the gate to the pool area, approached the side of the pool, and executed my strategy. The exhilaration of water, temperature, and movement was just as I expected. What I had not expected was that my swimsuit top would fill with air (or something else) during this entry. The result was two perfectly formed sets of breasts. One bared to the world and the other perched atop the first. I was probably the only one who knew this happened, but the result of this water encounter was an addition to my entry strategy, that of holding my top in place.*

And another:

> Cele, Adolescent, Fire~Box Campfire Memory: *I had been a Brownie Scout before I was a Girl Scout. Camp in the summer was one of my favorite activities. To be outside, to get to canoe, swim, ride horses, shoot arrows, and learn to exist in the wilderness was my idea of heaven. I loved studying and trying to catch those beautiful blue-tailed lizards whose tails would snap off so you couldn't catch them. The easily tamed ring-necked garter snakes were equally as lovable. But the most wonderful thing, by far, was the way those knowledgeable women built the campfires.*

At our camp, there was something called the Order of the Arrow, which was composed predominantly of counselors and advanced campers. It was an honor to be one of those. They were the ones who built the campfires. Now these were not ordinary campfires; they were built layer by layer, with the large logs on the bottom, in a box shape. Collecting just the right wood, piece by piece, took at least an hour. The wood of the box decreased in size until very tender wood was placed in a teepee shape on top. If all went as it should, one match would light the whole fire. The teepee would fall to the next layer, and so on, until the fire was a beautiful campfire. The smells of the wood were equally as wonderful.

In retrospect, I think the fire probably became a symbol for the wonderful rituals of camp. The wonderful times, friends, and beautiful voices raised in song all are blurred in this memory of fire.

Women and men who hear Sue's account of losing the top of her swimsuit and its floating plastic inserts immediately see her situation, a visual image that associates the event with a strong emotional response. The listener's embarrassment, identifying with the subject, is immediate and often is expressed as laughter (or at least a smile). We understand the importance of the social context of this memory intuitively. It depends on our knowledge of the stereotypic sexual importance of women's breasts and the social pressure to present them as being as large as is credibly possible. It also depends upon the presence of a perceiving observer. The swimmer's imagining what others might see creates one salient aspect of the memory.

The campfire as recounted was a complex sensory event focused primarily by vision for most of us and including many observations in nature. For Cele the campfire was interesting from several vantage points that depend on sensory input and highlight the importance of experience in the woods for this woman. We learn the centrality of the campfire as a socially important event, *"the wonderful times, friends,"* learning from *"those knowledgeable women."* We also see the campfire as an arena for learning about the particularities of wood and its uses. We found in discussion that the campfire was an experience shared by most of the memory-work group.

This pair of memories points up differences in the telling, the narrative of memories. Abram (1996) suggests that language "secretes" a translucent perceptual barrier to the subtle and rich differences that we can see, hear, touch, smell, and taste. The language used by the narrator strongly influences comprehension and engagement in the listener. In group analysis of the individual descriptions of memory, one goal was shared understanding. Repetition of words that are not descriptively specific, such as *"wonderful,"* *"beautiful,"* *"very nice,"* *"smells good,"* remind the reader of the tone of the memory without providing strong sensuous access.

The language of telling may be a veil that creates distance or it may pull us into participation in the story. The first memory is told as an unfolding event, the second as a description of a situation, a snapshot. A part of the hearer's identification with the young woman in the first memory springs from being drawn into how she felt. The unfolding of the story-pleasure in wearing the suit, anticipation of being with friends, planned entry into the water, jumping in, feeling the excitement and the water itself-draws the reader into understanding the interacting physical and social aspects of the situation and sets her up for the surprise appearance of two sets of breasts. In contrast, we learn about many parts of camp that Cele enjoyed in a sort of panoramic view, which leads into a description of building the campfire itself, the focus of the memory. In this narrative she is not an active participant in building the fire. She pulls the reader into the event more indirectly than does Sue in her account of the perils of entering the pool.

Hearing

In analysis, we also identified hearing as a single sense that stood out in several memories. The tie between the senses and social signaling is powerfully illustrated in relation to hearing and the sound of words. Most social interactions depend heavily on hearing and interpretation of speech. Reflecting on and abstracting from subtle intonations provide meaning that guides our responses to other people. Embodiment of language in speaking and hearing and the importance of perceiving meaning from visual cues that accompany speaking underscore the importance of sensuous experience in inferring social meaning. The anger or urgency in a parent's voice, the sneer that accompanies a teacher's pointed comments, empathy perceived in a friend's posture and tone are examples of the complexities of embodiment that inform us. The miscommunications many of us experience with e-mail may spring in part from the lack of bodily cues it provides.

As illustrated by the following memories, hearing may be central to personal science as well, here, in learning about the nature of air. In these accounts, the experience is inextricably woven with social interactions.

Anna, Child, Air~Breathing Memory: *One late summer afternoon, Anna, at a later preschool age, and her toddler sister, Katy Beth, found Dad in one of his usual resting places on the floor near the kitchen counter. He had been out in the fields during the day, making hay, and would come in for something to drink and a short nap. Dad could nap anywhere and one favorite afternoon spot was on the floor. He stretched out with his head on the bottom drawer of the cabinets that were handmade by his father, Grandpa Nelson, and identical to the cabinetry on the altar at the small country church. Those cabinets were also made by Anna's Grandpa Nelson, who died the year Anna was born. Katy Beth and Anna believed that their Dad lay on the floor to play with them or nap with them. So, as the girls rested their heads on their dad's chest they were astonished by the sound of his breath, in and out with rhythm and harmony to the heartbeat. It sounded like the air had to go a long, long way to or through Dad's body, and it would come out warm.* "Was this the same air that went into Dad? Is the air going in the mouth and out the nose, or what? Does Katy Beth's air do the same things, and can I hear it if I listen to her chest?" *The game of playing with the sound and touch of air continued.*

Sound takes on a crucial natural warning role in the following memory of a tornado.

Cele, Adolescent, Air~Tornado Memory: *We often went to the cellar when we were growing up. In southeastern Kansas, you just planned on severe weather in the spring. We got rain, wind, and occasionally hail. On this early summer evening, though, things seemed different. Mother awakened us right after the news and said we needed to go to the basement. Sleepily, we trudged grudgingly down the two flights of stairs into the basement. The rain began to pick up and the wind began to blow. Then, a train started towards the house. But there were no tracks nearby. What was that terrible deafening noise? It got louder and louder, the house trembled and shook. We sank closer and closer to the corner, all of us jammed in the little corner. Then it was over. My dad immediately said he had to go, but we were not to leave the house. Mother kindly let us go to the kitchen and stay with her while we began to get damage reports. My closest family friend's house took a direct hit; she was so frightened she crawled under the bed, thereby avoiding the glass window, which crashed where she would have been sleeping. Her terror in not being able to open her bedroom door after the storm still is very real. Her family was as panicked as she was.*

Fortunately, my dad [a physician] was able to give us injury reports occasionally, so we did not have to worry until the next day, like most of the town. No one was killed, but everyone took tornadoes a lot more seriously then; the sirens were put up and the citizens' band radio volunteer group for weather watching was founded within that year.

In Anna's memory of air as her father's breath, we find a series of questions. The answers to these questions are not explicit in the memory, though there is a sense that they are answered in the context of this experience as part of a game. In the game the breath observed is very closely tied to the cue element, air, as is the game focused on the sense of hearing. Here, the social context—air literally embodied in the children's father—is closely connected to a developing personal science. These children are learning about both the sound of air and the way it feels and are raising questions about whether it is the same air going into the body and coming out of it. The comparison of adult air and toddler air further extends the learning.

The intensity of the sound in Cele's memory is the backdrop for the actual coming of the tornado and the loss and suffering one fears may follow in its wake. The sensuous experience of the terrifying wind throws into relief the departure of the physician father to help others and the protection of the mother. Natural disasters heighten our senses, our attention, and both block and hone memory. In this case, the physical behavior and effects of the tornado are learned both by direct apprehension in the basement and by understanding the behavior of a close friend.

Additionally, the memory demonstrates the importance of physical reaction to the natural sound. It also demonstrates a way in which we begin to alter our intimate connections with nature. In the interest of safety, the community provides technological warnings to alert people of danger. As a result, we have become distanced from the signs of the coming natural event. We still notice the signs, but we probably are not as vigilant in listening to the wind and watching the skies as we were before we learned that the sirens would warn us.

Other sounds described in memories may not be central to the theme of the memory but are lures to the element itself. Consider the clunk of shovel on rock in an earth memory, the whoosh of the hot air balloon as the propane burner is lit, and the poof of igniting kerosene on a wood fire. Not surprisingly these all are described onomatopoetically. In contrast with overused words and turns of phrases, these have the potential to carry the listener to the sound itself, directly communicating about the natural world and breaking through language or what Abram (1996) refers to as the veil of perception secreted by language. The listener then assigns the significance. The physical embodiment of experience in a word is something that is similar across cultures. Words that sound like the sounds they represent are recognizable in any language.

Touch

Touch, which depends upon contact with our skin, seems very close to the elements themselves. Even though the experience is usually immediate in the large number of our memories involving touch, it may also be tacit as in descriptions of activities for which the hearers have their own experiences to draw from. For example, Cele, in a childhood memory, mentioned rubbing a calf's back, which conjures up the feeling of rubbing coarse hair for those who have done it themselves.

Phenomenologically, extreme reactions provide richer meaning than do neutral reactions in the memories and their interpretation. In the following memory, disgust with touching the element was prominent, and the depth of disgust was compounded by the inevitability of having to deal with rotting leaves year after year. The interaction between Sue's father setting an unavoidable task for his daughter and her strong negative feeling for the leaves in all their predictably chaotic behaviors and states of decay reinforces a lack of enthusiasm for doing anything like this in the pursuit of natural science.

Sue, Adolescent, Tree~Dirty Leaves Memory: *Every Saturday I was afforded the opportunity to "work" for my father around the house doing whatever chores needed tending to. They varied greatly, but two specifically occurred in the fall and left the notion of trees and their products as something less than desirable.*

The first was the leaves themselves. Our lot was wooded with oaks, both pin oaks and the ones that have bigger leaves. When the leaves died and fell, an ocean of leaves arrived overnight. My job was to rake them into piles. There were mountains of leaves, uncontrollable leaves and dirty leaves, wet and buggy.

The second dealt with the back steps to our basement. I was given the wonderful job of sweeping the steps, cleaning them out each fall and spring. Leaves and other unmentionable items would get blown down the steps and "grow" during the fall and spring. I will never forget starting at the top of the steps, where some light and air reached the leaves, allowing them to dry and respond to the winds. I would then sweep down subsequent steps. Each step held more unmentionable items that were wetter and "grew" more creatures and critters than I wanted to experience, ever. The bottom steps were the worst, and I worked without gloves, having to touch the "things" that hide among the leaves and debris.

Typically my strategy for dealing with these tasks was to put them off as long as possible, then I would work as quickly as possible, touching things as little as possible. My father showed no mercy, ever. Or at least that is my memory.

Although the possibility of Sue developing a personal science of decomposition might seem unlikely, she, in fact, has. She recognizes that light and air at the top of the steps keep the leaves dry and not very buggy, whereas the dark and wetter bottom steps are prime habitat for rot and a variety of unmentionable growing creatures. The effects of socialization in this case are obvious; Sue displays a reaction often reinforced in girls and in this way participates in distancing herself from this aspect of the natural world. She has learned quite enough about decomposition.

Although touch depends on close physical contact, it provides information well beyond bodily limits. The following two memories report experiences of natural phenomena rather different from the rotting leaves memory.

Bell, Adult, Earth~Earthquake Memory: *She was at Synergy that day teaching with the high school kids—probably the history class. They always sat at a long oblong table in that low one-room building at the top of the hill. She was discussing something when suddenly she felt a piece of earth roll under her feet. It was unlike anything she had ever felt before. She likely would have just ignored it, but it continued to happen. The earth was very distinctly rolling up under her feet, so much so that her legs and knees were pushed upward. She looked at the class; they were all staring at her waiting for her to do something. There was that moment of puzzlement—"What is this?"—then that moment of recognition, but it was an odd sort of recognition because this was like nothing she had ever felt before. You didn't feel an earthquake so distinctly under your feet like that. Then there was the moment of shock and that wondering if this was going to be the big one. Without a word, they all seemed to get up at the same time and run out of the building. You weren't supposed to run out of buildings, she thought, but it seemed so instinctive. For all of them, it was the natural thing to do.*

And:

Anna, Child, Earth~Muck Memory: *Anna was swimming with her sister Katy Beth at Spring Lake. It was a time when Grandma and Grandpa still lived there, before Uncle Bob and Aunt Jill bought the homesteaded property. The dock was an important marker — one side for swimming and the other side for launching small boats. The swimming side had sand dumped into it, defining the texture of the earth for the swimmers.*

Once again. Anna and Katy Beth started on the sandy edge, moving toward the end of the dock, the arbitrary point which served as the limit for the distance the young girls were allowed to venture. The adult suggestion was followed: walking out to the end of the dock and swimming back toward shore. Silently this time, Anna and Katy Beth recalled the feel of the sand, to the mixture of sand and smooth rock, to the mixture of smooth rock and rough, larger rocks. Soon after, the larger rocks were combining with silt and weeds, until the limit was reached: marked not only by the dock but by the MUCK!

In these two memories the difference in scale is marked. In Bell's earthquake memory the very earth beneath their feet moves and affects the classroom and beyond. In contrast, Anna and Katy Beth had individual experiences of the serial change of substrate along a stable layer of earth. The earthquake is the result of dynamic tectonic events, and the change in the lake bottom is spatial and relatively stable. Nevertheless, in both these instances the rememberer has learned about important qualities of earth. The young teacher at the Synergy School is initially puzzled by what she feels; the confrontation with the element is physical, intellectual, and social. Students wait for her to act, but who knows how an earthquake feels until it is felt, much less what to do in response to it? Her understanding eventually surfaces with theirs and they all react together instinctively.

In contrast, Anna and her sister have encountered the element in this context repeatedly. Even though Anna exclaims about the muck, the implied distaste for it does not stop these girls from intentionally encountering it in exploring the lake bottom to the full limit of their permitted space. Anna and Katy Beth expect the observations they make; they are a translation of the safety rules set by adults. In this instance, little girls exploring with their feet are adept at observing, comparing, and communicating to each other nonverbally, by posture and facial expression, the unseen substrate changes under the water.

It is important to note that these small girls were acting within long-term family traditions of playing, working, and washing in the lake, traditions tied to cultural heritage. The cultural/social setting for this experience facilitated and encouraged the development of personal science. Memory-work made us aware of the importance of such sensory exploration alloyed with socialization in its development. Anna and Katy Beth, as well as Bell in the earthquake, experience personal science, which relies on engaging with the natural world. Engagement is often physical, sensuous.

Engagement may also occur in traditional science. In the following memory Rae is engaged by the sensuous aspect as well as the intellectual.

Rae, Adult, Water~Quarry Pond Memory: *Rae, 21, is working on her honors research project in an old quarry pond outside of town. She loves coming to this pond by herself. The granite walls are steep but have convenient ledges for getting down to the water's surface. It's always damp and cool here, and she loves the smell of the brownish water where fallen leaves, among other things, rot. It feels good when the sun strikes the wall where she works. She likes knowing the names of some of the insects that swim here and enjoys operating the sampling gear and using the field chemical kit. She makes frames to hold microscope slides where algae will settle down to live and anchors them on ledges and other rocks that are under the water. After they are colonized, she can measure their rate of photosynthesis and then preserve them to identify and count the cells. This project is her idea and her execution, though she has a good faculty advisor who is helpful. She goes out to the pond again and again and again over her senior year. This day in winter she uses an axe to cut a hole in the ice so that she can retrieve some slides and put new ones out. It seems to take forever to get through the ice, and it's hard to manipulate things with such cold hands. The water seems to move like molasses through the hose from the sampler, and the reagents that fix the dissolved oxygen samples are slower yet. Nevertheless, this is her idea of a wonderful way to spend an afternoon: out, messing in water, doing things she knows she can do.*

This memory reflects a setting of traditional science in which both the technical observations and the pleasure afforded by several sense modalities are obvious. Solitary mental processes abound and are tied to sensuous experience. Furthermore, the ritual of the science learned in school has been well rehearsed by Rae, who is confident in this science. Whether the interaction of the learner is with a teacher, a group of learners, or alone, it constitutes social construction. In this case Rae's interaction with the equipment, the method, and the disciplinary framework is socially constructed, illustrating the classic Vygotskian idea that the individual is inseparable from the cultural (Vygotsky, 1981).

Smell and Taste

Smell in our world rides on air, and we associate taste with what comes in contact with our tongues. Olfactory stimuli such as burning leaves or baking bread can be powerfully evocative. These fragrances, like things we taste, carry us back to former places and events and often call up memory. Within our narrated memories, however, taste and smell were rarely described as central to experience, with such exceptions as the taste of dirt in Cele's early dirt potatoes memory. In contrast, in other memories, e.g., of a campfire or the fair, the aroma of burning wood or cooking hot-dogs is background to the experience recalled.

Nevertheless, the association of smell with the quality of air was important to a few memories, such as this one, in which the importance of socialization in the old-est child of the family is clear.

> Anna, Adolescent, Air~Bogeyman Memory: *The air was hot and humid and musky under the blanket. Anna knew that the exposure to the outside would be dra-matically different, but movement would draw attention to the made-up source of fear. Many weeks ago, Anna's sister Katy Beth had moved downstairs because the furnace couldn't keep up with the demands of the frigidly cold below-zero days and nights. This was the year the upstairs of the farmhouse was converted for the two ad-olescent girls, but the furnace was not yet converted from the wood and coal routine in the basement. Anna thought the hallway of the upstairs was a cozy place for her new twin bed with its loyal-to-the-Falcons purple and white bedspread. This way she didn't have to sleep in the same room as Katy Beth. But, at 14 or 15, the oldest and most responsible still had the nightmares of someone else in the hall. Tucking the heavy quilt made by grandma over her head provided a safe but smothering retreat from the fear. The sickish feeling in Anna's stomach may be coming from breathing the used air. Take a breath!*

In the following memory, the rememberer's close relationship to the element is linked to taste. Here the girls demonstrate personal science in their knowledge that the "real treats" are underground.

> Cele, Child, Earth~Dirt Potatoes Memory: *It was a late June day. Cele and Sarah were outside playing on the lush green lawn. As they wandered down by the old stable, they remembered the garden beside it. Sneakily, so they would not be seen by anyone working in the kitchen, they each pulled up a radish or two. Since they were afraid to prime the pump to get water to wash them (too noisy!), they simply dusted them off and ate them. Next they tried the carrots. Slowly, the two girls sneaked into the potato patch and dug the soft, rich brown earth to get the real treats. Again, they dusted the new potatoes and ate them raw. They were so good that even the rich brown dirt simply added to their delicious flavor.*

Although memories in which a single sense is highlighted initially may provide access to the experience for a particular listener or reader, a single sense cannot be divorced from the multiple modalities and meaning of the experience. Salience often comes from a multitude of sensory inputs and their integration. As von Hornbostel (1927, p. 83) commented on the interactions among the senses, "It matters not through which sense I realize that in the dark I have blundered into a pigsty."

Development of Personal Science

In this chapter we have commented upon memories that we believe illustrate personal science. We argue that personal science is embodied, that is, located personally in our bodies, and it has roots in our sensuous experiences as children. Our bodies as mediums for experiencing the world are anchored, as Carol Bigwood (1993) maintains, in "an interconnected web of relations with the human and the nonhuman, cultural and the natural" (p. 55). Within that web, girls develop complex relationships to nature and fruitful approaches to their science from an early age, as is particularly evident in the following three memories.

> Sue, Child, Earth~Sandbox Memory: *It is warm outside. Sue has just left the house by the back door. The screened door slams shut. She squints at the sun and hurries to the sandbox. Her neighbor and friend is there too. They are both wearing shorts, tennis shoes, and short-sleeved shirts. The play items in the sandbox are pails, spoons, shovels (miniature size), and an older metal tea set with plates and cups. The top of the sand is hot and dry, not uncomfortable, but definitely not wonderfully cool and wet like the sand that is down deeper in the sandbox. The difference between the two, the top and the bottom, is dissipated as they both mix the sand from the top and the sand from the bottom together. However, no matter how much bottom sand they mix with the top sand, it is never quite wet enough to "play with."*
>
> *Sue and her friend need the wetter sand to play successfully. Sue can solve the problem because the sprinkler is going on the lawn. She carefully, very carefully, piles sand on a plate and places it strategically on the grass within the sprinkler's shower path. Her father grunts disapproval at this solution. He demands that she take the plate back to the sandbox. He then moves the sprinkler closer to the sandbox and indicates that sand must remain in the sandbox. Sue then piles sand on the plate and places it on the wooden rim of the sandbox. This move enhances the chances of getting the sand wet because the edge of the sandbox is closer to the source of water, the sprinkler, than the box itself. Also, the sand is still "in the sandbox" by her definition. Her father did not like this strategy, and kicked the plate back into the sandbox, getting sand in both their faces. He then picked her up by the arm and dragged her into the house, spanking her with every step along the way.*

Here Sue is well aware of the temperature differences in the sand from surface to the depths of the sandbox. She also learns the degree to which the sand must be wet to be used for her work with her friend. Her observations along those two gradients are central to her efforts to obtain more water from the sprinkler. Her misunderstanding, intentional or not, of her father's view of the limits of the sandbox precipitates his anger and the end of her pleasure in comparing, mixing, testing the properties of sand and water. In this memory personal science is embedded in family conflict; exploration of the world of texture is cut off by an angry parent enforcing his rules.

Rae, Child Earth~Digging Memory: *Rae is perhaps 4 or 5, digging a hole under the spiny locust tree. She digs and digs a hole around herself with a trowel. She likes the smell of the damp, dark soil and the way it feels in her hands. She wishes the hole were deep enough to be over her head. She thinks this would be very nice since her brothers are digging a very deep hole for a hideout over in the empty lot, and they will not let her go into it. But her excavation gets deeper very slowly. She abandons her own digging; it's too hard.*

Maybe a day or two later, Sam and Fred take her to visit their hideout. They lead the way through the dead leaves with a lantern, or was it a flashlight? She has to crawl on her belly through a narrow tunnel down into their dugout. She sits with her back against a dirt wall and looks up at the ceiling that holds up the dirt overhead. The room under the earth seems both small and very large and only barely connected to the rest of her world.

The story is told with descriptions of the body. She smells the dampness, feels the dirt, its texture and water content. Its density prevents her from accomplishing what her brothers are doing. The control for observation and comparison is her own work. Although there is no meter for light, no tool for calibrating soil density, Rae explored and knew her own situation, made decisions, and appreciated her brothers' invitation, in part, because of the comparison. She admires the hollow space in the ground created by them. The earth looms overhead, cramps her breath perhaps, and darkens her vision, but comparison with the earlier personal experience provides the context for appreciation of the space.

Bell, Child, Air~Arm in Air Memory: *She's sitting in the front seat next to her father. It's a hot summer day and they're on the turnpike, maybe driving to Boston or maybe going to the beach. Her fingers are hanging on the bottom edge of the window, and she starts to lift them, and she feels the force of the air. If they were more flexible, they would bend backwards. Then she moves her hand into the wind force, and she can allow it to be picked up and tossed by the air. Soon she's got her whole arm out there, and she gets totally lost in allowing her limb to be played by the air. There's an odd sense of disembodiment as if she's watching someone else's arm and the air could take it away. She starts pacing with the rhythm and moves her arm down toward the sideview mirror. She feels the force under the arm moving very fast, up and down starting at her fingers and moving to her elbows. She applies some resistance to let her arm ride the wave, and then she picks up her hand slightly and spreads her fingers, and the gust forces her arm abruptly back, feeling a letting go—close to a sense of flying. The spell is somehow broken and she's back in the car and back to being bored.*

Bell clearly describes her experimentation with the air and the engagement with the memory by the hearer rides on the tension between embodiment and disembodiment in the experience. Although described from an adult perspective, the experimenting with different movements, positions, and angles dominates this memory and marks it as a girl's personal science.

In this chapter we have explored the importance of attention to sensuous experience in analyzing our memories. Here we have uncovered the role of the physical in our experiencing nature, both human and nonhuman. In contrast, the focus on the cognitive and social in trying to understand the development of individuals often overshadows the physical dimension of learning. We can see why embodiment is an important feminist theme with an emphasis on the conscious study of women's bodies and society's views of their bodies. Yet, Birke (1999) suggests that although feminist theory has emphasized the "process of cultural inscription *on* the body, and . . . the cultural representation *of* the body, the body that appeared in this new theory seems to be disembodied—or at the very least disembowelled [*sic*]." She adds, "theory, it seems, is only skin deep" (p. 2). Few feminist thinkers have extended their ideas about, for example, the uses of sensory access of the body in constructing knowledge. Scott (1997) speaks to this kind of limitation: "Although standpoint theory uses the fact of embodiment to argue for the authority of feminist knowledge, its theorists have been almost as reluctant as other feminists to draw directly on bodily knowledge" (p. 115).

We must ask, what is the role of the physical body in the development of women's relation to nature and their science? Jagger (1989) has offered one avenue for knowledge emerging from the body in her conceptualization of "outlaw emotions." She asserts that feelings are a kind of bodily knowledge. An outlaw emotion, such as uncontrolled laughing at an authority figure, is socially inappropriate to a given situation because it is at odds with the values of the dominating figure. Outlaw emotions tell us about the implicit values of the prevailing ideology and through them, Jagger has helped us recognize the importance of the body in knowing. Maynard (1997) suggests that knowledge of the importance of the body can be used to deconstruct the conventional understanding of science, emphasizing Scott's (1997) work. The objective gaze characteristic of conventional science has permeated the biological representation of sensory experience as relatively passive, our bodies simply receiving the stimuli of the outer world (Birke, 1999). Our work underscores the active interplay among sensory experience, exploration of nature, and the evolution of personal science.

When we began memory-work research, we were focused on potential explanations/descriptions from cognitive and social standpoints. In analysis we initially were surprised to find that the sensuous dimension in the memories offered fruitful access to understanding. Although we accepted the unity of the social, cognitive, and physical, we were slow to recognize the contribution of the physical to understanding. In this chapter, we have situated our knowledge in our sensory experience, based on an awareness that emerged in memory-work. Sensory experience is an important component of science-whether personal or professional.

As a result of examining the role of the sensuous in the construction of memory, we have rediscovered and remembered moments when we connected our selves physically in a kind of personal science, a way of knowing or making sense of our world. As mentioned in chapter 1, some have suggested that experience in nature provides children a feeling of "continuity with natural processes" (Cobb, 1959, p. 538). We suggest that this sense of continuity grows out of the closeness to the natural world which many children experience and which informs their knowledge of nature. This closeness gives rise to spontaneous concepts that children develop about the world (Vygotsky, 1986). Distance from physical experience in nature increases with growth and, we suspect, contributes to our forgetting earlier close connections. In fact transcripts of our analysis show the difficulty of recall. Some in this collective found it much more difficult to recall memories of adulthood that were associated directly with the elements than memories of childhood. Their adult memories focused on the metaphorical role of water, for example, rather than its more direct sensuous aspect, which was more common among childhood memories.

Phenomenology provides a useful perspective for understanding our connections to the natural world. Gibson (1986) proposed the term affordance to represent what the environment "offers to the animal . . . either for good or ill" (p. 127). We have discovered that memories contain evidence of how physical/sensory experience is inextricably bound up with the meaning we make of the environment. This is affordance, the idea that meaning arises in the fusion between a physical experience and the context in which it occurs. Awareness of affordance may be key to identifying personal science, as in Sue's sandbox memory in which an affordance of sand is its potential changes of state and therefore its usefulness in construction. The meaning of sand for Sue cannot be separated from the physical experience rooted in the sense of touch. Similarly, in Anna's muck memory, the changes in substance under the girls' feet offer a physical "lived distance" (Cataldi, 1993) within the boundaries of safety. At the same time, the experience of comparing those familiar substrates provides an affordance of personal science, learning about the textures and particle sizes of substrates in a systematic way. For both Sue and Anna, the meaning of their experiences with the earth was simultaneously mental and physically sensuous.

We conclude that personal science, in contrast to traditional science, is anything but a disembodied, objective undertaking walled off from human interactions, a finding consistent with Bigwood (1993), Freire (1985), and Hubbard's (1988) suggestions. Our memories suggest a range of social settings where personal science has taken place. We have highlighted exchanges that value curiosity and observation of nature as well as those that discourage further exploration of nature, particularly its physical and biological components.

The next chapter carries our search for clues to the development of women's relationships to nature and science into the realm of metaphor. One path toward understanding, that is, toward making sense, lies in joining the metaphorical with our observations of embodied perception. This path extends the range of sensory experience, i.e., making sense, into making meaning using metaphor.

CHAPTER 5

Metaphor: Girls in Their Elements

It is as though the ability to comprehend experience through metaphor were a sense, like seeing or touching or hearing, with metaphors providing the only ways to perceive and experience much of the world. Metaphor is as much a part of our functioning as our sense of touch, and as precious.

—Lakoff & Johnson, 1980, p. 239

As suggested above, metaphor is integral to our lives, a notion that has been eloquently and repeatedly asserted (e.g., Petrie, 1979). Among the better known presentations is Lakoff and Johnson's (1980) accessible analysis. Although others maintain that metaphor is not so basic to understanding (e.g, Green, 1979), Lakoff and Johnson believe that metaphor is central to our conceptualization of experience.

The familiar role of metaphor in literature has been exhaustively analyzed in literary criticism, particularly that dealing with poetry (e.g., Lakoff & Turner, 1989). However, Lakoff and Johnson remind us that metaphor is not just a matter of language as many assume; metaphor structures the development and understanding of concepts. For example, many of our fundamental metaphors are based on physical orientation, particularly of the human body. Conceivably, up and down are learned as valueless directions, and, as a result of our physical experience in health and illness and life and death, they have become the metaphors "up is good, down is bad" (Lakoff & Johnson, 1980, p. 17). Other metaphors provide ways to think about intangible ideas and emotions as entities, for example, "a wall of hate," or "a well of generosity." More complex yet are structural metaphors that may use one metaphorically based concept to structure another. For example, in introductory biology texts (e.g., Audesirk & Audesirk, 1996), information is said to be "coded" in the chemical composition of DNA, a biochemical metaphor derived from using codes for communication. DNA composition (its structure) is eventually "translated" into the chemical composition of protein, providing another biochemical metaphor. This second metaphor of communication depends on the conceptualization of the first.

The metaphors we choose and their meanings vary from culture to culture, and, of course, individual perceptions of metaphors vary enormously. As a group, although we accepted metaphor's fundamental place in understanding, we initially resisted examining its role in the meaning of memories, our socialization, and our relationship in nature. Nevertheless, a strength of memory-work is the deconstruction that can take place in group discussion of memories and their metaphorical meanings, though that work is not easy. Many metaphors have become such conventional, common turns of phrase that their metaphorical nature is difficult to recognize. Their apparent transparency is a barrier to our seeing one impact: Metaphor's everyday role in socialization. We see through metaphors; we look through the glass pane of metaphor, not seeing its function. At the same time we see experience through (i.e., by means of) them. For example, metaphors that have to do with order are used routinely to "order" girls, to socialize them, to prevent their becoming "disorderly." Haug (1987) points out the links between hair and order, the unacceptability of having wild hair, unruly hair, the hair that suggests the untamed or sexually uninhibited nature of the witch. Girls are stereotypically expected to stay clean, dirt indicating a disorderly nature. Similarly, clothing must be tidy, buttoned, zipped, and tied. The implications for keeping sexuality under wraps, under control, in order, are clear. We, too, learned that by analyzing metaphor, we uncovered clues to our socialization.

In this chapter we analyze metaphors in our memories as an approach to understanding our relation to nature. We also analyze use of the elements, ostensibly cues from nonhuman nature, to evoke memories that speak to the issues of fear, efficacy, relationship, growth, and development. Some metaphors draw us in close connection to particular aspects of nature. Some push us away. Examining metaphors makes clearer how distance from nonhuman nature develops and suggests implications for our interactions with science in school and perhaps with traditional science.

We begin with our struggle as researchers to understand the uses of several levels of metaphorical analysis. We then proceed to issues raised by our memories, discussing the findings that emerged from analysis of metaphor in the memories associated with each of the elements. Finally we consider some of the cultural connections to metaphors that are important in learning about nature and traditional science and suggest that this analysis is useful for thinking about the science we learned in school.

The Uses of Metaphor

Initially we had divergent beliefs about how to examine metaphor specifically. Bell preferred a more direct cognitive analysis than imaginal/metaphorical representation.

However, in discussion, she said she was "constantly grasping for a way to organize this stuff. It drives me crazy to muck around without having closure" (8/25/95, p. 19). This comment, not immediately apparent as a pun and perhaps metaphorical, was part of a discussion of Anna's memories, in which lake muck appeared (muck memory). Metaphor often came up in our analysis. For example, Bell speaks of grounded theory and, looking for expansion, Rae asks, "What does that mean?" Bell replies, "That means that the data will generate the theory." Rae: "Grounded theory—rising from the soil. I like that. . . . The spring of knowledge—the spring of theory" (2/8/95, p. 11).

Our attempts to talk about the nature of memory itself in metaphorical terms repeatedly broke down or were foreshortened. In seeking an image for memory, onions, sediments, and kernels were repeatedly raised for discussion. Talk tended to center on the specific details of onions, kernels, and sediments rather than on their usefulness in understanding the nature of memory.

Nevertheless, we came to agree with Schratz and Walker (1995), who in memory-work research "found metaphor particularly productive, not just in identifying metaphors within the text but in having the group develop the practice of using their own metaphors to understand it . . . to ask, 'What does this remind you of? What pictures come to mind?'" (p. 48). Eventually we asked: "What do metaphors tell us about socialization in connection with the elements?" At one level, it is clear that the use of metaphors from nature, even from the elements themselves, is probably common in expressing feelings. Fire is particularly telling in the metaphors that speak to its danger. We think about "putting his feet to the fire," we say that "the fat is in the fire"; we watch for signs that family or colleagues will "burn out" and "look for the smoke" coming out of a parent's ears. We wonder whether we have gone "from the frying pan into the fire."

At another level, the telling of the story, how the memory itself was narrated, may provide metaphorical meaning. A metaphor may be provided by putting two disparate experiences together within the memory as in the following example cued by fire.

Bell, Adolescent, Fire~Sweaty Palms Memory. *The burner serves as a magnet for all the negative emotions surrounding that class. She had heard that the teacher, Mr. Wilson, terrorized his students. She never knew when he would call on her to perform a piece of an activity or answer a question, and god help her if she didn't do it just right or have the correct answer for his question. He needled the students, embarrassed them, maliciously teased them—unless of course it was a student he liked, and then he displayed a sort of kindness that was belied by his unpredictability. You never knew when he would explode at one student or all of them. . . . There's also a sense that she was unsure of the Bunsen burners. She was not quite sure how they worked or how they were lit—she had to relearn how to work with one each time she had to light it—and each relearning made her vulnerable to a possible attack by Mr. Wilson. The next year she had heard that Mr. Wilson's college-age son had committed suicide.*

Bell's sweaty palms memory is told with strong negative emotion. The teacher *"needled . . . embarrassed . . . maliciously teased"* the students who did not perform this task perfectly. Every relearning made Bell vulnerable to *"attack"* by this teacher who might *"explode"* at any time. The story ends with a sentence that reveals that this teacher suffered as well as created suffering. Discussion of this memory also brought to the surface our tacit recognition that the story provided a kind of metaphor for the loss of girls from the sciences, which, we believe, may be a common consequence of teachers treating them in ways perceived as harassment in the science classroom.

Not only were metaphors present in the memories; the group used metaphor to help provide insight. For example, in Bell's childhood tree house memory below, Bell's feigned excruciating pain, which she thought a boy might experience, upon falling from her brother's tree house to straddle a rope below. This was interpreted by the collective as a gendered ticket to membership in this group of boys. She had hoped to gain acceptance in his group, though when that failed, she reported screaming at her brother and his friends above in the tree house.

Bell, Child, Tree~Tree House Memory: *The boys were inside the tree house and she wanted to join them. They were all at least four years older than her 7 or 8 years. They didn't want her up there at all, and she wanted to be included. She ignored their taunts and exclamations that she wasn't allowed and started to climb up. All of a sudden she felt a push from someone, maybe it was Frank Lawrence. She landed on her feet but straddling a rope that had been strung between two trees. She wasn't hurt very badly at all—a slight pressure or maybe even a rope burn on her crotch. However she pretended that she was in a lot of pain—almost the same sort of pain a boy would feel after getting hit in the groin. She might have even forced a tear or two. She remembers screaming at them about what they did to her, but they were unconcerned.*

In other cases, the metaphorical meaning that came with group discussion transformed understanding of both the rememberer and the group. A discussion of the nature of trees as kinship associated with commonplace representations of the tree of life and family trees provided the context for learning. Rae came to see that her affinity for trees can be thought of as a personal relationship to trees. The exchange within our group clarified the connection between Rae's solitary nature and her sense of family. That is, as a collective we came to understand that she, in pursuing experience apparently alone in nature, turned to trees as an extension of family.

Metaphors served yet broader uses as well for us. They stimulated thought (as will be seen in our analysis of the glass images associated with Bell's adolescent tree memory) and encouraged a change in perspective (as shown in the discussion of Rae's trees). Another function of metaphor was as a safe outlet for divergent views within the group. For example, in the exchange between Bell and Rae about grounded theory, Rae used a pair of metaphors both to check her understanding of Bell's statement and to explore other possible meanings. Metaphors also promoted a healthy resistance to premature closure. We sometimes dealt with our uncertainty about findings by offering metaphors that described the difficulties of the analysis and kept the door open to further discussion. For example, when we saw that our attempts to graph our data were a result of being seduced by the quantitative or, to use Porter's (1995) phrase "trust in numbers," we turned to other approaches.

Making Meaning

In chapter 4, "Making Sense," we argued that memory-work provided evidence of the direct experience of nature which children, particularly, may have. One type of knowledge that springs from direct sensuous experience is a kind of personal science. In this chapter, we describe how metaphors associated with the elements illuminate basic human experience. For example, fire represents through metaphor, warmth in relationship ("rekindle the fire"), and connection to generations of energy and love ("home fires"). Conversely, it can signal anger ("a fiery glance") or emergency ("put out the fires").

In analyzing memories and our discussions of them, we extracted metaphorical meaning and organized various interpretations by element (earth, air, fire, and water). In doing so, we recognized the multivalent nature of the classical elements and their metaphorical and mythic connections, which were reflected in our experience. Recognizing these connections allowed us to discover the emergent themes of fear, efficacy, relationship, and growth and development.

Fear

Although all elements provide essentials for life, they also wield forces that can harm us. Fear appeared in memories that described direct contact with an element or indirect experience associated with an element. Fear also was associated with the metaphorical extensions of a variety of memories.

The terrifying side of the elements lurked in the shadows, mostly unremarked upon, for all of us. Still, more than half of us reported memories that involved natural disasters; natural disasters are memorable. For example, Cele's tornado memory deals with destruction and concern for the safety of friends and family in the path of the storm.

Water may also threaten life. Torrential floods, precipitous waterfalls, raging rivers, and towering waves stir an instinctive fear of water. As one of us said, "I have always been in awe of fire, more so of water . . . almost frightened of water" (12/12/95, p. 11). The danger of water was clearly implied in a childhood memory cued by air. Here the focus for Cele was the challenge of swimming underwater across a pool, then once again at the surface *"gasping for air and enjoying that feeling of air, not water [flowing] into the lungs . . ."* (underwater swim memory). In discussion, we found several references to water's power to cover the unknown and then to reveal what has been hidden, both physically and metaphorically. Speaking of an adult water memory, Sue described part of her fear of water as lying in what she imagined was hidden by the flood. When the water receded and she saw what had been carried by the force and darkness of the water, it was no less frightening. Even apparently clean water may be contaminated and its hazards unrecognized as remembered in Cele's childhood experience: *"Cele tasted the cool, clear-flowing water from the stream. Her father loudly scolded, 'Don't drink this water, even though it looks clear, it comes largely from a mine run-off. It will make you very sick'"* (stream dam memory). Water could be crystal clear and at the same time harbor toxins.

Several other childhood memories illustrate the attendant dangers of air and fears associated with loss of control in air, such as the fear of falling from high in the air or anxiety over breathing, getting enough oxygen (memories of Sue and Rae). Like Crawford et al. (1992), we found that loss of control was often associated with fear.

We learn at an early age that fire is associated with danger. Cele's early memory of running across a hot floor furnace demonstrated with clarity her knowledge of the danger of heat. This memory metaphorically combined fear, danger, caution, and, at the same time, the attraction of risk and the warmth of family care, *"The small girl ignored the pain; after all, she was about to get the longed-for chocolate milk. She hurriedly grabbed her robe and ran back to her grandfather"* (chocolate milk memory). Fire was the cue to unite her early knowledge of danger and thrill with warmth of love and nurturance.

Earth has the power to suffocate us. The association of earth with death, fear, and the underworld was evident among our memories as a childhood fear of being buried alive. Reading a newspaper account of moving graves stimulated a discussion of the possibility and horror of being buried alive by mistake, the heart of the childhood memory recounted by Bell (moving the graves memory). We also were immediately moved by her account of an earthquake experienced as an adult, which brought to the fore the possibility that earth in motion results in destruction and death.

As we commented in discussing Haug's account of memories associated with order and hair, dirt is often invoked to control girls. In an adolescent earth memory,

Cele had been warned about the dangers presented by chat, the waste from a mining operation. A more subtle and complex meaning was hidden in the metaphors of the following memory.

> Cele, Adolescent, Earth~Chat Piles Memory: *Although our parents had repeatedly warned my brother and me (and later my foster sister) not to ever play or park around the chat piles, it was only in adolescence that the reasons for their seemingly unwarranted fear became apparent. I grew up in a lead and zinc mining area of southeastern Kansas. Chat is the small, crushed waste rock that is brought to the surface as lead and zinc are mined.*
>
> *We always assumed that dangerous people, derelicts, and drunks, hung around these small rock mountains; that was why we were not to go there. One morning, our father said we had to go see something. He loaded us into the car; we drove approximately a mile south of town, where one of the more visible chat piles was. But it was no longer there! Half of it was missing. The dirt county road that bordered on the south was also gone. What was left was a large gaping, and growing hole. It was only then that I remember questioning my dad, the local physician, about the people who had lost their lives when they fell into the hidden shafts on the chat piles or suffocated when the chat swallowed them up. A similar scare occurred about two years later, when much of the town of Pilcher began to sink. The whole area, from Miami to Joplin to Pittsburg is sitting upon mine tunnels. Some of these have filled with very polluted alkaline water from the earlier mining days. Others are just waiting for those occasional earthquakes to dramatically alter the landscape of this part of the country.*

The chat piles readily and unpredictably collapsed, and the account of this memory contains prohibitions about the people who might be encountered at the chat pile. Our discussion probed the use of the social prohibition, which controlled a girl's movement and reinforced earlier teachings of classism. The "dirty old men" are commonly invoked to promote fear and discourage girls' freedom to explore. Cele's brother, in contrast, was warned only of the physical danger, the possibility that a collapsing chat pile would crush his car. The literal physical danger was used to control his movement; his sister had to negotiate complex social interpretations of the chat pile as well.

Efficacy

Efficacy is a theme we found closely allied with fear and sometimes with danger. Will we succeed? Will we fail? Will we encounter danger? Many of our memories that dealt with these questions were cued by fire and water, two elements that are obviously necessary to our well-being and dangerous as well. Sometimes the danger is mediated by social control. For example, Bachelard (1964) in *The Psychoanalysis of Fire* discusses the cultural and historical values that shape the meaning of this phenomenon. Stemming from some of these values are childhood prohibitions against using fire. For this reason, when a child steals matches to head for the fields, or in Bell's case the front driveway, it is a "problem of clever disobedience" (p. 11).

> Bell, Child, Fire~Matches Memory: *She's got a book of matches and this might be the first time she has ever lit them, and what she wants to do is light every one. She's feeling nervous about being caught because she is not supposed to play with matches. She's also wondering whether Frances, the nosy, gossipy, and incessantly talky woman who lives next door, will catch her and tell her mother. Her mother always seemed to find out about all of the stuff Bell wasn't supposed to do. Bell lights the matches—one by one, although she's drawn to lighting the whole book all at one time. She's not sure what would happen if she did that—whether the whole thing would blow up and burn her hand and singe her eyes. There is a distinct image of the gray burned matches all lying in a pile and barely distinguishable from the gray driveway.*

The lighting of a full book of matches one by one, as remembered by Bell, was her secret attempt to gain the personal knowledge of fire that those older than she possessed. This desire to know more about fire than fathers, mothers, or teachers is termed by Bachelard, the Prometheus complex, named for the Greek titan who gave fire and the arts to humanity. We might add that transgressing the prohibition of fire is also based on a desire to know what these people do not want us to know from our own personal experience, our developing personal science. Bell is successful in achieving this knowledge. We will discuss this memory and the issue of clever disobedience further in chapter 7, "Family Landscapes in Nature."

In the memories written about adolescence and young adulthood, the meaning of fire appeared to transcend the polarities of danger and warmth associated with this element to encompass ritual deeply embedded in the natural phenomenon. Adolescence brought memories of fire that were more personal and socially more complex than those remembered from childhood, entailing such issues as performance, identity, and social justice. Bell described her feeling of success as an adolescent in meeting the challenge of building a "matchless fire." *"That was the most important thing—using only one match"* (matchless fire memory).

In Cele's adolescent box campfire memory, she described the wood used in a campfire. She intended to write "tinder," the literal meaning for the nature of the combustible material. What she wrote, however, was *"very tender wood."* An allusion to adolescent girls was the effect of this inadvertent word choice. In our collective analysis, the tenderness of the wood was metaphoric for the uncertainty and vulnerability of girls at adolescence. Their concerns about efficacy often make girls fearful.

Competence was tested in water in several adolescent memories, and the outcome was hidden in the sense that it was not predictable. Often associated with the test was our fear that we would be found wanting. Anna's learning that water can be "*hard*" occurred in the context of trying to ski in the midst of other family members who were very competent at waterskiing (waterskiing memory). Would she be able to stay up? Who knew until she was "in the swim?" Bell was literally tested on her swimming and life-saving skills in waters controlled and judged by an intimidating martinet of a teacher. Her uncertainty of her competence was partly hidden by the turbulence that the act of demonstrating her swimming skills brought to the surface of the water. In this part of the test her strokes were executed smoothly and well, reassuring her. However,

> Bell, Adolescent, Water~WSI Memory: *Then it was time to demonstrate our lifesaving skills. He [the instructor] slowly paired them up—each one dreading whom they might be paired with. They all wanted someone light or at least someone buoyant. She knew she was weak on this one—and he knew her vulnerability. He paired her with the lead weight in the class. The one who sank to the bottom of the pool if he didn't do anything to stay afloat. Her stomach wrapped into an even tighter knot.*

She believes that she has been set up for failure. Indeed, water lets her down; she does not have the strength to lift her test partner from the pool. At the same time she came to know that the disagreeable instructor could not allow an incompetent lifesaver to pass; water can kill.

In another memory, water and efficacy are tied together in a demonstration for a biology class.

> Rae, Adolescent, Water~Carrot Water Memory: *She is trying to ease a piece of glass tubing into the top/shoulder of a carrot, but it's hard to do. The carrot does not easily accommodate the tube, and she is afraid that if she forces it, the glass will break. The glass is cloudy with the carrot debris on her hands. She's sweating, her sweater (ha!) clammy on her hot body. She really wants this to work both to see it herself and because she wants it to be right for Miss Schwartz, the biology teacher. She is not at all certain it will work.*

Rae sweated over whether water would rise in the glass tube when she set up this demonstration. Would her felt incompetence be hidden by the movement of water or would it be revealed to a much-respected teacher by water remaining motionless in the beaker? "*She returns hours later to find that the water has climbed up the tube and she is relieved and pleased.*" Efficacy will be discussed further in terms of mastery in chapter 8, "Power of Girls."

Relationship

The importance of human relationship in nature is a theme that emerged from metaphors associated with the memories. In several cases, the narrative of the memory was presented against the backdrop of water in air, i.e., mist or fog, which was metaphorically important to understanding the relationships. Water in its vapor state may obscure the clarity of details as we saw in discussion of Anna's childhood memory of the traditional Finnish sauna.

> Anna, Child, Water~Sauna Memory: *The memory is a vivid recall of a sauna, which is usually conducted in extended family units on Saturday. Somehow, it doesn't seem like a Saturday, but details are gone. Every sense is involved as the water sizzles into steam on the hot rocks and envelops naked skin. Details are hazy.*

Anna spoke of "*details* [that] *are hazy*" in recall and commented that "it's as if there are these sort of bright lights on pieces of the experience" (9/13/95, p. 3). She conveys the limited visibility of a memory that nevertheless may be recognized as central to understanding socialization. She communicated clearly to us the cultural importance of this ritual. We learned about the construction of her role in the matriarchy of her family from group discussion of the memory.

Metaphors also linked the space between the known and the unknown in aspects of relationship. These metaphors were often associated with memories cued by air. Particularly striking is the fog described in this adult memory of a meeting between friends, both on the verge of significant changes in their lives.

> Cele, Adult, Air~Fog Memory: *Cele decided to take a break from studying. . . . Todd was lounging around the [residence hall front] desk waiting for something to do. Cele said she was ready for a break and asked if he would like to go for a walk. It was an unusual July night for Lawrence, Kansas, since it was extremely foggy. Besides holding the cold air several hundred feet up, the warm air held large, suspended water droplets. There was essentially no traffic, since it was impossible to see more than a few feet in front of you. . . . They decided to walk to the duck pond near the center of campus. As they walked, they talked about their futures. Cele was due to be married in less than a month; Todd was to graduate in December. Both were feeling more than just a little bit odd at the affection they had for each other in spite of their individual plans. Finally, behind Strong Hall, they got to the subject. Cele noted that she wished she had more time with Todd; Todd wished that she were not already committed. There didn't seem anything else to say — in spite of the darkness and the fog; they chose to go their separate ways back to the dorm and beyond.*

On this occasion a metaphorical fog hid the futures of Cele and Todd. In the discussion, she recalled being struck by the risk taken in being out in such thick fog and in speaking of the road not taken with this friend. Neither the separate futures for these two friends nor the future as partners that might have been is known. The known is that their decisions have been made; for now their paths have been chosen.

This memory also illustrates metaphorically the fog that often fell over our analysis. Group discussion of this memory documented, once again, questions we as feminists logically might have asked and did not. Cele was about to marry someone whose affinities to her were not clear while Todd, the man in the fog, was a fellow student with whom she had spent a good deal of time and for whom she felt considerable affection. How did these relationships affect or not affect her professional plans, her professional development? We never asked.

The tree as an image for family, for the branching of development, the implications of roots in lineage and ethnicity also arose in discussion of relationships. We justified the choice of tree as a cue for memories on the basis of its being alive in contrast to the other elemental cues, which are not. However, tree's suitability may also lie in its interactions with air, earth, water, and fire (as potential fuel). All vegetation, indeed all life, integrates these elements.

Tree is the cue for a memory of striking contrasts for Bell. She remembers her drive as an adolescent through the wooded countryside to work at a glass factory. The factory itself, on a site from which all trees had been removed, was landscaped with green-tinted glass. This barren, "*stark and desolate*" vision throws into sharp relief the absence of nonhuman nature. Here innovative technology provides a disappointing synthetic substitute for the natural world.

Bell, Adolescent, Tree~Glass Factory Memory: *When she was in high school, she worked part-time for Sentry Industries, the place where her father worked as transportation manager; the place that was owned by her brother's soon-to-be-father-in-law. She worked there on Saturdays answering the phone and writing up orders for glass. The factory was newly built and located in Webster, a small town just outside of Worcester. You had to take local roads to get to it, so the trip could take anywhere from half an hour to 45 minutes. It was a beautiful route through lots of country and wooded area. The first time she saw the new factory she was struck by how stark and desolate it was—they had totally stripped it of surrounding trees. It was surrounded by parking lots, and in particular the slopes surrounding the lots were landscaped with broken, green-tinted tempered glass.*

In the discussion of the glass factory memory, we learned both the amazing properties of the glass in this factory and the properties of this family, most of whom worked here. Sheets of the glass could be walked on, the surprising strength and safety underscoring the possibilities that technology conjures up. Metaphorical analysis of Bell's memory made clearer to the group the intricacies of family relationships that had been opaque to us, and many other metaphors came up in discussion as we came to see their potential impact on the socialization of girls. Testing the bow of the glass sheet, its physical tolerance, brought to mind testing the limits of that intangible—family tolerance. Walking on glass reminded us of the difficulty of knowing what you can and cannot say in a family. We again recognized that metaphors provide ways to think about intangible ideas and emotions as entities.

In other ways, a tree proved a salient image. In Cele's adolescent experience, a specific tree presented metaphorically the real possibility of her death and represented her relationship with this tree and her concern for their mutual survival. When the tree clearly was dying a few years later, a new understanding of mutuality came home to Cele. Here the woody record of history also served as a metaphor for the transition to adulthood.

Cele, Adolescent, Tree~Car Tree Memory: *I couldn't believe it. The tree was dying. Impossible; this was the tree that was so important in my adolescence. I was certain the small, strong tree would outlive me by many years. Over the years I drove past it when I went back and forth to college and graduate school; I watched it mature into a gorgeous shade tree, with only a couple of noticeable flaws.*

This tree and I crossed paths right after I turned 16. It was a glorious day, the last day of summer school. I knew I had done well in my classes, another pair of A's. I had ridden with my older brother and his friend in the wonderful 56 Chevy hardtop; I always rode in the back during the 30-mile ride. The luxury of a full-month vacation was upon me. The National Girl Scout round-up started in less than a week; I was one of two elected to go from our district. What fun!

Then, nothing. Later, I was told that the man in the old car had pulled out in front of us, and Gary had tried to make the mile road. We slid into the tree, with the force of the impact at the left back-seat door. The tree and I both ended up with scars to match. The eighty-eight stitches sewed my skin back together; the tree was not so lucky. That scar from 1959 was to be with it forever. It was a small scar in 1959, but later it grew to at least two feet across and three feet in height. My broken pelvis and spleen would heal, but the tree carried the "tilt" from the force of the car throughout its remaining years. It took me a long time to get the courage to go see the tree, much less get out and touch it and see the remains of the car still strewn around its trunk.

Over the subsequent years, we sustained a kinship that I have enjoyed with few other objects. It was much more than a landmark; it was a part of me. It was as if those birds and other creatures shared my home in that tree. For years after 1959, I watched it mature; I worried when the road was repaired that it would be cut. Not ever did I expect it to succumb before I did. Losing it was losing a part of me. . . .

Growth and Development

Many of the memories included references to planting gardens, harvesting vegetables, and the work associated with cultivating the earth, both appreciated and disliked. Although we associate growth with life, growth as a metaphor extended to the appearance of hated objects to be cleaned up year after year by adolescents. Bell wrote of *"those boulders [that] grew back"* in her mother's garden every spring. *"Why did she have to do the dirty work before her mother started the planting?"* (growing rocks memory). In Sue's adolescent tree memory, dead leaves blew down the back steps to the basement where they rotted and *"grew more creatures and critters than I wanted to experience, ever"* (dirty leaves memory). She too regarded this as dirty work and hated it. For Sue, earth was primarily dirty, a property that, in her memories, earth shared with every other element except air. This generalization on the prominence of dirt extended to her name for the collective, "The Dirt Group," as we learned when her husband referred to us thus in conversation with a colleague. At the other end of the spectrum, Cele as a child happily discovered that the flavor of potatoes eaten straight from the garden was enhanced by, made *"delicious"* by, dirt. For Rae, planting a garden was both a literal physical experience and a metaphor in her adult memory cued by earth.

> Rae, Adult, Earth~Family Planting Memory: *It's April and Rae, just [turned] 30, is in the backyard watching Eric planting, with Dawn (3) and Lara (2) helping. Grace (8 mon.) is probably asleep in the house. The soil, not as red here as it was before many gardens were planted over the years, smells good where Eric has turned it and prepared the bed. Dawn and Lara take turns filling coffee cans from the hose and a tub and carry each can of water to Eric who is soaking each hole after he places a plant in the soil. Rae enjoys their mutual pleasure in the task and her own sense of satisfaction in facilitating it.*

Discussion highlighted her facilitating the efforts of her preschool daughters to help their father plant flowers, her pleasure in this kind of growing a family.

In adolescent efforts to grow and mature, conflicts between should and should not are common. A memory of smoking cued by air is a typical instance of new experience: experimenting and risk-taking. The description of sensation: "My *head was at the top of the 15-foot room and my legs and body were still seated on the bed. What an experience. I was so high."* Here the meaning of the memory can be derived from the metaphor of being *"up in the air . . . an expanse of air all around me"* (Sue, smoking memory). As Lakoff and Johnson (1980) maintain, up is good, and in this instance the vertical orientation of Sue's body, her head feeling as if it was at the ceiling, provided the high.

Speaking of the "vertical life of trees," Bachelard (1988) observed that these plants serve the imagination with respect to all the classical elements. He reminds us of the impact upon the imagination of fierce wind in the branches, the hidden lives of roots beneath the earth, the rising of sap and its connection with water, and the immanence of fire implied in the woody trunk. All of these—fierce wind, the hidden, rising sap, and fiery moods—are associated with adolescence.

In our discussion, verticality itself also emerged metaphorically in connection with a young adult memory of being simultaneously physically high in a tree, high on acid, and high on mind and the search for new cognitive spaces.

Bell, Adult, Tree~Up in a Tree Memory: *The time seemed to drag on inordinately waiting for . . . the pizza. Bell decided that she would assume the post of lookout for the pizza "lackey." She went out the front door and climbed the tree just in front of the house. It seemed as though no one knew she had left. Sitting up in the tree she didn't get the sense of a novel space or a new sensation to occupy her mind. She felt lonely and began to wonder what she would do if the pizza "lackey" arrived. A bit of paranoia began to set in. She quickly climbed down from the tree and returned to the safety of the house.*

The complex structure of our adult fire memories was linked to our self-development and fulfillment. Some were as simple as expressing the desire to re-member to open the damper when lighting a fire in the fireplace (Sue). A campfire with friends and family concluded "*a perfect day*" for one rememberer (Cele family fire memory). By another fire, contentment lay in listening to the howling coyotes and thinking about the potential of a baby soon to be born (Rae). Most striking was one member of the group's identification with a candlelight vigil. "*Thousands were marching with candles, and it was absolutely silent and completely dark except for the thousands of lights moving down the street. It was compelling and deeply moving for her*" (Bell candlelight vigil memory). Among the thousands of lights and thou-sands of people, identity issues were illuminated and possibilities opened for this woman.

Rituals often mark important life passages. Throughout our memories, as evi-denced in the candlelight vigil memory and several from childhood and adoles-cence, fire serves a ritual function. Lakoff and Johnson (1980) note that we create rituals all the time. They maintain that it is through ritual that metaphors become experiential, providing us a personal identity as we practice over and over the en-actment of a cultural belief or value.

The metaphors we live by, whether cultural or personal, are partially preserved in rit-ual. . . . There can be no culture without ritual. . . . Just as our personal metaphors are not random but form systems coherent with our personalities, so our personal rituals are not random but are coherent with our view of the world. (pp. 234–235)

Many of our memories of both adolescence and adulthood are characterized by ritual. The analysis of fire memories brought the ritual of the campfire forward in our understanding of our connection to nature. Some remembered the procedure for starting and lighting a campfire; some remembered the ritual of looking for the wood; some remembered the talk, chocolate-marshmallow treats, or stories told around the fire. Each of these memories cued by the element fire represented some important aspect of the ritual.

Examination of fire memories uncovered the rituals that reinforced power and mastery in our adolescence. Here efficacy was linked in the campfire memories to a strong connection to older, competent women (Cele box campfire memory), competence in lighting the fire (Bell matchless fire memory) and the satisfaction of connection among peers sitting around the camp's ritual fire (Sue campfire memory). Cele was explicit. Fire itself for her "was secondary to the songs, the camaraderie, the feel" (1/9/95, p. 10).

Science in school presented repeated challenges to our efficacy. Many of these also represent ritual as seen in several adolescent memories cued by fire. Both Bell's and Rae's memories of lighting a Bunsen burner portrayed the adolescent anxiety attendant on completing this ritual successfully. Bell evocatively described the *"gradually drying wet marks made by her sweating palms"* on the black slate lab table (sweaty palms memory). The need to perform the rites of school science properly may loom so large that the content meaning of the work is lost, and only the social pressures to measure up in the classroom are learned. Anxiety over lighting the Bunsen burner correctly seems to have stood in the way of understanding both the principles of maintaining a useful kind of fire with the burner and the meaning of the experiment itself for Bell.

The Myth of Distance from the Elements

The idea that we are separate from nature is a cultural myth. We said at the outset of this chapter that although we accepted the role of metaphor as basic to making meaning, we had resisted examining its role early in our analysis of memories. In this chapter we have explored metaphors used in the narrative of memory and more broadly the metaphoric content of whole memories. This analysis dovetails with examining the metaphoric meaning of the elements themselves, which is reflected in myth. Although myth may distance us from the elements in its narrative representation of them, myth also provides fundamental understanding of the role of the elements in human experience. Though we rarely think consciously about the mythical construction of the elements, the metaphoric grounding in myth underlies our language and experience of nature in the Western world. For example,

> Fire is the ultra-living element. It is intimate and it is universal. It lives in our heart. It lives in the sky. It rises from the depths of the substance and offers itself with the warmth of love. Or it can go back down into the substance and hide there, latent and pent-up, like hate and vengeance. Among all phenomena, it is really the only one to which there can be so definitely attributed the opposing values of good and evil. It shines in Paradise. It burns in Hell. (Bachelard, 1964, p. 7)

The metaphors we have encountered in our descriptions and analysis of memories cued by the elements bring with them the baggage of mythic associations.

Cultural expectations for the relationship of girls and women to nature also color metaphors that touch on the elements, their everyday meanings and uses. Stereotypically, girls are not supposed to "get dirty"; "earthy" has both negative and positive implications. Girls may be told that they are "out of their depth," a metaphor implying that they would not be at ease, or perhaps safe, in particular waters or intellectual discussions. If they develop interests in the interactions in nature, girls will be encouraged to "understand the mechanisms" because nature is widely believed to be understood best as a machine. By examining such metaphors both in our memories and our analysis, we have come to understand how socialization can shape our relationship to the natural world. We recognize, for example, how we might be taught not to get our hands dirty, which may keep us distant from nature. This teaching springs in part from concern for women's conventional behavior. The conventional emphasis on objectivity in the conduct of science also reinforces distance.

We have seen through memory-work how our own metaphors for ourselves in relation to nature reflect and perhaps have shaped our development. The close metaphorical association of water with sustaining life and growth is perhaps clearest in Rae's memories. She is absorbed by the role of water in life and by the life in water. From a fascination with water in play as a child to delight in learning of the underwater habits of dragonflies to achieving competence in experimenting with algae under the ice, all evidenced in memories cued by water, she came to associate water with wonder and intellectual growth. More broadly she claimed, "water is a metaphor that I feel is . . . fundamental in my life" (11/3/95, p. 13). In water, she is in her element. For the rest of us, our experiences with water sometimes interfered with our understanding this kind of relationship to an element. As one of us remarked, "All of my water would be in a cup or in a kitchen" (11/2/94, p.13). Our own experiences may make it easier or more difficult to understand experiences of another. Eventually, however, the meaning of water for Rae was uncovered by group discussion.

Some lives are closely tied to an element, and others are not. This observation and our knowledge that memories in which experience is closely connected to elements are clustered in childhood led us to wonder about how distance from nonhuman nature develops and is imposed.

When we see our metaphors more clearly, we are in a better position to change our views of ourselves and therefore, as Bateson (1991) suggests, to change the way we view the world. As we said in the preceding chapter, we have learned to revalue the sensuous and the physical roots of our connections with nature as a result of this work. We agree with Bigwood's (1993) assertion that it is more important to experience nature than merely to record it. Considering the sensuous led us to see the importance of personal science. One result of our thinking about metaphor and the distancing of women and girls from nature is a clearer understanding of the roles that science in school may play in that distancing.

Science in School

The role of metaphor in science itself has been extensively analyzed for traditional science (e.g., Kuhn, 1979; Garfield, 1986). We, in contrast, have focused on metaphor's role in what we have called personal science. A number of memories examined in chapter 4, "Making Sense," illustrate the importance of metaphor in our understanding of personal science as, for example, does Cele's glacier memory of her adult encounter with the "*talking, living beast*" of a glacier. In her account, the metaphor gave her a new understanding of the motion and power of glaciers.

Recognizing both the sensuous and metaphor in the course of analysis led us to recognize affordance, the fusion of physically based experience and context in making meaning. We have spoken about personal science as an affordance of physically based sensory experience. The sensory experience cannot be separated from the comparisons, measurements, and generalizations that we construct about the world. Neither can it be separated from the social aspects of our experience. Similarly, the quality of experience is an affordance of the rituals of science in school.

The anxiety of girls performing the tasks of laboratory science reflects the indivisibility of the demands of the setting and their emotional reaction to it. That affordance, anxiety, may be a source of motivation or a deterrent to further scientific curiosity and exploration. For example, Rae's experience with the Bunsen burner had a positive affordance even in the face of anxiety.

Rae, Adolescent, Fire~Bunsen Burner Memory: *Rae, 17, breathes the acrid air of the high school chemistry lab. The pale green walls are bare except for the Periodic Table and three dirty windows, and the black lab benches, too, are nearly bare except for a Bunsen burner and the sparker to light it at each place (or was it every other?). She wonders whether she will be able to strike a spark fast enough after opening the gas valve to light the stream of gas at the burner before it accumulates and explodes in her face. She scratches [the sparker] as fast as she can, lights the gas, and relaxes enough to breathe again.*

Still sweating, she turns the notched wheel at the base of the Bunsen burner with her right thumb and forefinger to adjust the amount of air the flame is permitted. She is very happy to see the oval cone of blue at the base of her flame that tells her that she has the proper mixture of gas and air. She finds the shape and the blue color of the inner flame pleasing, as pleasing as her success in producing it.

In contrast, as we have seen, for Bell lighting the Bunsen burner at school meant anxiety that entailed humiliation, shame, and dread (sweaty palms memory). The strength of these feelings, the lived depth of this experience (Cataldi, 1993), constituted a negative affordance. Bell was never adequately at ease in such a laboratory learning environment to pursue her curiosity about nonhuman nature.

A positive school science experience may lead to wider curiosity about nature. Notice in the following memory that a quite conventional lecture on the importance of chemistry was transformed for Anna into something full of wonder by a demonstration.

Anna, Adolescent, Fire~Magnesium Memory: *The [chemistry] class demonstrations were particularly motivating, perhaps because of the novelty of instructional method. One demonstration was particularly exciting. The instructor, Mr. Olson, was discussing the need to know about chemistry and the chemical properties that supported us in everyday life. At some point in the lecture, he dug around in his cupboard and pulled out a strip of aluminum-colored material that he lit. The material burned with such a brilliance that it appeared supernatural. Anna's fascination with the material was lingering as she continued to listen to Mr. Olson. At some point in the semester, extra materials were given to students for independent use. Anna managed to get an envelope of magnesium and took it home to demonstrate the wonder of the burning of this material to her sisters and others who would watch. Anna kept the magnesium in her jewelry box.*

For Anna, fascination with the element was the result of coupling an ordinary lecture with the burning demonstration. Magnesium came to have many different meanings for her: it was mystical; it was teaching opportunity; it was a treasure. It sparked multiple possibilities for learning and was part of her abiding interest in earth science. This memory is important to our understanding of making new meaning for ourselves in nature, the subject of the next chapter.

Although we expected that memories of science in school would be readily cued by the elements and that these memories would provide insight into our socialization as apprentice scientists, we were surprised that we did not find this to be true.

Developmental dilemmas, issues of competency, and self-consciousness dominated memories of science in school, particularly in adolescence. This is entirely consistent with the developmental literature. In some instances school science was ultimately inviting (Rae's carrot water and Bunsen burner memories) and in other cases, it quite effectively dampened interest in natural science among members of the group (Bell's sweaty palms memory). We have seen that desire to succeed and fear of failure are signal emotions in adolescence, and we predict they strongly influence continuing interest and achievement in exploring nature and traditional science. Hermanowicz (1998), commenting on his study of astronomers, underscored the importance of ambition and self-doubt in achievement in science. Self-doubt "fuels ambition on the one hand while threatening and at times killing it on the other" (p. 190).

CHAPTER 6

Making New Meaning: Creative Acts

It is in playing and only in playing that the individual child or adult is able to be creative and to use the whole personality, and it is only in being creative that the individual discovers the self.

—Winnicott, 1971, p. 54

Creativity, for us, is making new meaning. New meaning is often expressed in an art form using pen, sound, movement, paint, or clay. For us memory-work created a new understanding of our relation to nature. We discovered how we made new meaning in nature when we played, solved problems, imagined, explored, and experimented. Acts of play in nature throughout our lives seemed to bring new values and differing perspectives to everyday life. We pursued this observation and coded all memories for play and creativity. The subsequent analysis is the basis of this chapter.

We debated whether the focus of the chapter should be creativity or play. The two are closely intertwined as Winnicott claims and as we saw in our analysis. We asked, "what does creativity in play have to do with science?" Traditional science often requires distance from its objects of study, the natural world. At the same time, creativity is thought to be essential to science as it seeks knowledge of nature, and the process of science includes the creative work of asking new questions, using imagination, exploration, experimentation and solving problems. Memory-work enabled us to see that creativity in our play as children reinforced close connections with nature. Those connections, particularly with non-human nature, had been obscured and in several instances lost in our socialization, yet connection undergirds our interest in nature and, for some of us, in science. We identified and reclaimed the creative impulses evident in our memories, and this approach opened our awareness that new meaning in nature can emerge through play. This chapter explores these new meanings within the boundaries of the physical and social spaces in which creativity can exist. By examining these boundaries in our memories and analyses, we learned how creativity is hindered or facilitated in our lives.

Creative Spaces

Boundaries may be physical, psychological, emotional or social; typically, they are socially constructed and intensely interwoven. We are particularly interested in the affordances associated with the creative spaces these enclose. Cataldi (1993) speaks of lived distance as space. We extend that meaning to the space where the creative acts of play occur, indeed where all creativity occurs. Here affordances of anxiety, joy, curiosity, boredom, and freedom arise.

Positive affordances that facilitate creativity can be garnered from social interaction. These may include zest, action, knowledge, worth, and a desire for more connection (Miller & Stiver, 1997). For example, Cele's watermelon memory illustrates the possibility of the creative outcome within a powerful social context.

Cele, Adolescent, Water~Watermelon Memory: *It was another in a long series of stunning summer evenings. My friends and I all had planned a late summer party to take advantage of one of these evenings. It was a quadruple date; four of us who hung out together and our dates. The plan was to go to Joplin, get a watermelon, and then go to the park and eat it. Naturally, we had no silverware, but that was never a problem.*

Our first stop was the Farmers' Market in Joplin. It took up a whole city block and was populated by rows of vendors. We strolled the various booths, looking for just the right watermelon. Finally, we found the dark green, huge Black Diamond that we knew had our name written on it. One of the guys plucked it up and we offered the vendor the $3 required. He said he could not take our money. We turned to another; he said he could not take it either. After offering approximately four vendors money for this wonderful melon, we just walked out with it. Naturally, we assumed the police were hot on our trail. As we made our way to the car, our consciences did not get any easier. We knew we were going to transport this particular watermelon across state lines. Wow! Did that make us felons?

Transport it we did to the little park where we often enjoyed the creek, swings, merry-go-round, and hot dogs. While that watermelon may not have been the best watermelon I ever had, it is the most memorable. To this day, all of us remember "stealing" that watermelon. It had to be one of the highlights of the summer after graduation.

Goldstein (1999) has emphasized the importance of the relational in making meaning, the social construction of knowledge. The social context energized Cele and her group to action and continuing connection. Questions of worth were raised by the activity. Here the boundaries include adolescent fears of being arrested for stealing, the summer evening, the group of eight friends, and the social structure of the Farmers' Market itself.

In this multidimensional creative space adolescents solve problems in their play together. This group has acquired a new understanding of the potential ambiguity of moral responses to multifaceted problems. Moving from this memory, more gener-ally, what are the connections between creativity and play?

Play

Play is not easily categorized. As Sutton-Smith (1997) notes, "We all play occasion-ally, and we all know what playing feels like. But when it comes to making theoreti-cal statements about what play is, we fall into silliness. There is little agreement among us, and much ambiguity" (p. 1).

Memory-work revealed the importance of play to us. Play is thought by some to be a childish waste of time with no connection to the serious work of the adults who may say of children, "they're just playing." According to this restrictive perspective, unstructured time, free play, is a luxury for the very young. However, as quoted at the beginning of this chapter, Winnicott (1971) asserts that play is at the heart of human experience. Within play at all ages all kinds of creativity can occur. We are particularly interested in what play teaches us about nature, in its many social and nonhuman facets.

For example, play may be hidden in the memory of an adult experience, as in Sue's memory.

Sue, Adult, Earth~Cat Prints Memory: *I was the first person up that morning, Sunday. I hurried out to the kitchen to start the coffee and turn up the heat. Each evening prior to bed I make an effort to clear and clean the counter top so that the morning is tidy and less irritating for all. My best efforts were thwarted. From one end of the counter to the other, at even intervals, were reddish-brown footprints. You could see the arrival and exit of Albert (or Peggy Sue) as (s)he strolled to the sink for a drink and then explored the rest of the cleared space. This bit of information let me know that it had been wet last night and that at least one of the cats had clearly sur-vived an evening out. But, I quickly sponged them away so that I would be the only one who knew.*

In this memory, Sue's work as a homemaker hides her creativity in figuring out what the cats were up to the night before and describing it as an engaging story. The play of her creativity was secret; she sponged the evidence away.

Is this truly play? We are not alone in wondering what qualifies. Sutton-Smith (1997) notes that play has been characterized in different ways and from multiple perspectives that he calls the rhetorics of play. For example, Nachmanovitch (1990) focuses in part on the "spontaneous . . . [and] disarming" qualities, which for him are associated with childhood; play may be childish. For him, growth over the years entails increase in complexity in a life. Thus play (a simple and childish enterprise) is increasingly difficult to engage in.

> There is an old Sanskrit word, *lîla*, which means play. Richer than our word, it means divine play, the play of creation, destruction, and re-creation, the folding and unfolding of the cosmos. *Lîla*, free and deep, is both the delight and enjoyment of this moment, and the play of God. It also means love.
>
> *Lîla* may be the simplest thing there is—spontaneous, childish, disarming. But as we grow and experience the complexities of life, it may also be the most difficult and hard-won achievement imaginable, and its coming to fruition is a kind of home-coming to our true selves. (Nachmanovitch, 1990, p.1)

Still, some have seen abandoning play as part of the transition to adulthood, playfulness having no role in the serious endeavors of adulthood as constructed by society.

> It may often have been true that a society was better off denying most curiosity and playfulness to prepare for a life that was necessarily both real and earnest, and that it has been sufficient for most individuals to live out their lives on the basis of a body of accepted cultural knowledge—indeed, the loss of playful curiosity may be the entry fee to adulthood. (Bateson, 2000, p. 192)

On the other hand, our assumption of childhood as simple and of adulthood as far more complicated perhaps reflects our societal construction of these times in life (Morss, 1996). Childhood may be, within a different constructive lens, more complex than our culture simplistically assumes. If so, we could narrate an entirely different story of childhood; in fact as little as 300 years ago, children were considered miniature adults (Degler, 1980). In short, according to the cultural story we now tell, play is more characteristic of childhood than it is of adulthood, and this was borne out in our memories across age.

We also observe that not all play results in an obviously creative product. Some appears to be satisfying and repetitive though not apparently involving a pursuit of new meaning. For example, one may ride a bike in a figure-eight pattern over and over again just for the physical pleasure of the act.

Still, some play depends on the imaginary and is associated with improvisation in the rhetoric of play literature (Sutton-Smith, 1997). Imagination and innovation, as important aspects of creativity, are valued in this view of play. Likewise, Vygotsky, suggesting the evolution of play over an individual life, saw its connection to adult creativity. He theorized how imagination transforms childhood play into a higher mental function for adolescents in fantasy and results in scientific and artistic creativity for adults (Smolucha, 1992).

We have chosen to recognize multiple and various indications of creativity in the memories we analyzed. Like play, creativity is multifaceted, and scholarly attention has focused on the creative person (Piirto, 1998), the creative product itself (Amabile, 1983), and the creative process from problem solving to spiritual journeys of life (Cameron, 1992). We have organized this chapter primarily around creative processes and the situations in which they occurred in our memories.

Our interest in the social and physical spaces in which creativity (and play) occurs has led us to see the importance of small geographic places. In these places, flexibility, freedom (Vygotsky, 1994), and psychological safety for play (van Oech, 1983) are often possible. The prominence of small places as a setting for creative play emerged from our analysis of childhood memories. In these spaces, we made connections with things animate and inanimate that took us beyond the reality for which we were being socialized. These memories sometimes offer a reality that is different from an adult perspective.

Rae, Child, Tree~Willow Tree Memory: *A long hill/lawn?/empty lot? is over the hill from her house. No one but Rae is there. Maybe 2/3 of the way up the slope of the hill is a very large willow tree. Rae plays under it, happy, hidden by the branches that come down to the ground. She has never seen anyone else on the hill; she has never seen anyone else under the tree. It feels like a wonderful place, secret and green, smelling like dirt and green.*

We see from the willow tree memory that Rae loves to play under this tree where she is surrounded by the leaves all around. She likes looking up and out through the leaves and plays by herself, pretending who knows what. She is just happy to be in this place, outside of time and the usual, a secret place *"smelling like dirt and green."* This small space offers an affordance of the creativity of pretending and playing. In the following excerpt from her digging memory, she experiences a new space.

Rae, Child, Earth~Digging Memory: *She has to crawl on her belly through a narrow tunnel down into their dugout. She sits with her back against a dirt wall and looks up at the ceiling that holds up the dirt overhead. The room under the earth seems both small and very large and only barely connected to the rest of her world.*

Nabhan and Trimble (1994) remind us that many children prefer small places as opposed to the large sweeping vistas of oceans or mountains often preferred by adults in their search for peace and beauty. Bachelard (1964) has suggested that the attraction to small spaces may be an attraction to nesting or we might suggest that it is the comfort of a space to a child's scale. Rae's memories also suggest that beyond the familiarity offered by this small space, she is experiencing a space that is different and separate from the world for which she is being socialized.

Emerging from memory-work is the importance of both the space and the processes of creativity. Within the creative space we found imagination, problem solving, exploration, and experimentation, all of which may contribute to new meaning.

Imagination

Creativity is linked to imagination in play and one type of imaginative activity is creating fantasy. In the following memory, Rae prepares a place for fairies.

> Rae, Child, Water~Fairy Rock Memory: *Rae is perhaps 5. She is playing alone under the forsythia bush that arches over a large, mostly flat rock. She covers a large part of the rock with pieces of moss to make it soft and puts water in the depression in the rock so that fairies can come in the night and swim in the water and lie down on the moss.*

Rae rewrote this memory, as described in chapter 3, and the transformation that appeared in the rewrite is evidence of a new perspective.

> Rae, Child, Water~Fairy Rock Memory Rewrite: *Rae is maybe 5. At one end of the rock garden is a forsythia bush, whose branches hang down almost to the ground. Under the bush is a large, mostly flat rock where she likes to play. She knows that fairies come out at night, and she works at fixing up the rock as a place where they might come then. From the damp places around the rock and other places in the yard she digs out the pieces of moss she needs to cover most of the rock so that it will be soft for them to lie on. She fills a depression in the rock with water so that they can swim there. Near the water she puts a few small sticks as a bench they can sit on.*
>
> *She has never seen the fairies, but she is certain that if she could be outside, hiding near the rock at just the right time in the night, she would see them and maybe could be with them too if only she were lots smaller, their size. Although she does this work entirely by herself, her brothers and parents know that she does it, and they neither laugh at her nor tease her about it and do not say that the fairies might not come. Sam even takes a picture of the fairy rock.*

Extended analysis of this memory resulted in greater detail and reflection in the rewrite, including consideration of what materials were needed for the fairies, where the materials came from, as well as the naming of the fairy rock. Rae provided more detail about what would be needed in her social interaction with the fairies: bench, right time of night, behaviors of fairies, being smaller so she could interact. She also added reflection about the family support for this imaginative play. The importance of this memory for Rae is evident in her statement that Sam's photo of the fairy rock hangs on her wall today "because it reminds me, when I have doubts, of a time when I was creative" (12/1/95, p. 34).

The boundaries of the creative space in this memory are both physical and social. The fairy rock itself was no larger than four square feet in area and was covered by the canopy of the forsythia bush. It offered a relatively private space and a rock with an irregular surface suitable for developing a small landscape. Rae's family supported the work; that support was an affordance of this creative space for her.

Rae's memory clearly incorporates a sense of possibility beyond the here and now. A central idea of this chapter is that the collective's experiences as children, early in our acculturation, illustrate that more possibilities were available. Rae's memory, though she did not actually see fairies, suggests her awareness of other ways of being, perhaps as a result of having been read to about fairies. It was a challenge for us to consider whether a child would consider interacting with fairies. In our discussion of this memory, she described her sense of fairies based on her childhood belief in them. "They were concrete; they made noise; they were female; they came at night. I believed in the fairies and wouldn't have made a place for them if I didn't. I would have loved to have been there at night with the fairies, but their size, not as a big person" (10/11/95, p. 12).

Some adults report seeing fairies. Hayward (1999) reports that the older population on the Isle of Skye off Scotland was interviewed, and they all said they saw fairies, "but when the education came the children were told fairies don't exist so they don't see them anymore" (quoted by Hayward, p. 69). The fact that Rae's is the only clear-cut memory of this kind may suggest that the rest of us were already taught not to see or remember, perhaps even consider such things as fairies. More generally, adolescent and adult memories in our work that demonstrated imagination are rare.

Exploration and Experimentation

Moments of imagination and wonder create striking cognitive and emotional experiences. Perhaps in contrast, exploration and experimentation are more obviously experiences of physical activity. In our memories of these moments we are creatively doing. Bell's early arm in air memory clearly conveys a playful experimentation or exploration of the element air. Similarly, Anna explores the characteristics of air with her father and sister in the breathing memory. Their activities lead to new understanding and a different awareness of air. The creative process brings new meaning to the girls playing.

We consider experiences like these to be personal science. In Bell's and Anna's examples, the discovery by the girls is creativity in the service of such a personal science. Vygotsky (Ayman-Nolley, 1992) theorizes how creativity develops the mental tools for scientific and artistic endeavors, demonstrating the link of imagination and cognition to professional science and the arts. Creativity and imagination were part and parcel of our personal science. They both were enhanced by and simultaneously facilitated our engagement in the natural world. We found that the skills and abilities gained from these imaginative and creative engagements in nature could be carried into and further refined in adulthood. As we look back in our own lives, we posit that creative personal science has opened up possibilities of competence beyond science in childhood and adulthood. When more possibilities exist, there are greater opportunities for people to experience competence and expand their sense of themselves.

Exploration and experimentation in adolescent and adulthood memories more closely resemble the problem solving of work. We see this in the efforts of Rae in her biology lab (carrot water memory) and the work of the single mother making a fire in her fireplace (Sue's fireplace flue memory) or erecting a Christmas tree (Sue's crooked Christmas tree memory). Less common were instances of more formal experimentation, such as Rae's quarry pond memory that detailed part of her college training as a biologist.

Problem Solving

We might ask, is all problem solving a form of creativity? Are some problems solved in routine and expected ways that we are hard pressed to see as particularly creative? Do others rely on new approaches for their solution? We would argue that a problem that is new to an individual requires some creativity in its solution. Alternatively, an old problem may be approached in a novel way, which is creative within the individual's experience.

Perhaps one of the more clear-cut examples of problem solving in our early memories is in Sue's sandbox memory. She sets up the sequence of the elements of a problem. The characteristics of the sand, its temperature, and the need for moisture are described. She is explicit, "*Sue can solve the problem because the sprinkler is going on the lawn.*" The memory details the added complexity of the problem with the difficulty presented by her father's rules. "*She . . . piles sand on a plate and places it strategically on the grass within the sprinkler's shower path.*" This solution is unacceptable to her father, so Sue "*then piles sand on the plate and places it on the wooden rim of the sandbox.*" The consequence of the second solution was further anger from her father. The problem-solving strategy met the criterion of the task; however, the boundary of her father's intervention cut short her enjoyment of the solution.

In adolescent play, many problems may present themselves. The overt intent of play for Cele as an adolescent in the watermelon memory was pleasure in the social group. The friends were looking for the perfect watermelon for the "*stunning summer evening.*" Their success in the intended problem-solving process brought about another problem.

The way in which the word "*naturally*" is used in that memory suggests how problem solving and imagination are occurring in the playful life of adolescents. "*Naturally,*" the adolescents will solve the problem of eating a large fruit without utensils, and just as naturally, they were self-conscious about transgressing the law as they imagined it.

Figuring things out also is a kind of problem solving as we saw in Sue's cat prints memory. There Sue figured out both what the cats had been up to in the night and how to prevent family discord over their walking on the counter. In the following adult memory, Cele's solution to a mystery is reached intuitively.

Cele, Adult, Tree~Allergies Memory: *Cele sits down in the warm spring sunshine to wait for the bus. In a moment or two, she notices that her eyes have begun to water and her nose has begun to drip. She can't imagine what is wrong; she doesn't have allergies. She glances around at the flowering trees and automatically moves away from them. Almost as quickly as she gets upwind from the trees, her eyes quit watering. Well, she used to be able to say that she didn't have allergies, at least.*

Here Cele is at first surprised that her eyes are watering and her nose is dripping. She does not know what has caused this reaction and automatically dismisses the possibility that she has allergies. Still in problem-solving mode, however, she looks around her, sees the flowering trees, and intuitively moves away upwind of them. When she sees the result of this move, she realizes that allergies may be the cause. Her immediate response seems to spring from knowing about trees and their pollen.

Problem solving may emerge in everyday creative tasks (Csikszentmihalyi, 1996). Rae finds great satisfaction in managing everyday tasks while flying with an infant.

Rae, Adult, Air~Flying Memory: *As the sun rises, Rae nurses Dawn and wraps her in a blanket on her lap. When breakfast is served, she rests the tray directly on top of her sleeping three-month-old baby and enjoys the sparkling snow below as she eats. She is happy that Dawn sleeps and that she can eat and look out the window, content with her cleverness as a mother.*

Cleverness in a small thing, an everyday task, is an example of what Csikszentmihalyi (1997) refers to as an autotelic activity, one that is done for the internal pleasure of doing it. For Rae it is her pride in her cleverness, her satisfaction in accomplishing her work as a mother and enjoyment of a peaceful moment.

Our exploration of creativity springs from an interest in our interactions in the natural world. We want to know how we have made new meaning in our connections to and within nature. Thus far in this chapter we have examined several facets of creative process that raise questions about the character of creative space. We turn now to two kinds of experience that appear to strengthen our ties to the natural world and, at the same time, facilitate creativity. In flow, time is suspended in a challenging experience that promotes creative endeavors for many people. Wonder, the second experience we address, is closely allied with fascination and appears to lead us to repeated interactions in which creative thinking and other activity can occur.

Flow

Flow is associated with creativity pursuits in which a person's intense practice leads to suspension of linear time, which is often pleasurable. In flow, a person faces a challenge and is intrinsically highly motivated to meet it, paving the way to optimal functioning (Csikszentimihalyi, 1996). Play, though not always creative, often involves "flow" in our memories and analysis. Play as flow may be situated in nature and within a sensuous relationship to the elements in childhood. Rae's memory of sifting dirt is a good example.

> Rae, Child, Earth~Alley Dirt Memory: *They sit playing in the soft, yellow dirt, tracing aimlessly, sometimes with a purpose that is not clear. She likes the soft dirt sifting through her hands, taking shape on the ground. There is no sense of time, just absorbing play with these children, with yellow-brown silt in the sun of the alley.*

In discussion, Rae commented "it was a suspended moment that was so pleasurable" (5/5/94, p. 5).

As Bell points out in her early arm in air memory, some activities do not have an end goal of solving a particular problem, but are highly engaging in the pleasure and performance of the activity (autotelic). She's moving her hand in the wind outside the car and learning about air. .

> Bell, Child, Air~Arm in Air Memory: *Soon she's got her whole arm out there and she gets totally lost in allowing her limb to be played by the air. . . . The gust forces her arm abruptly back, feeling a letting go-close to a sense of flying. The spell is somehow broken and she's back in the car and back to being bored.*

In contrast to Rae's sensuous pleasure in the yellow silt, Bell achieved flow with highly cognitive entailments and an awareness of her own mental tools (Vygotsky, 1990).

Lewis (1998) describes flow and the total absorption among children playing with butterflies.

> For the briefest time they seemed to be an example of something learned in which no part was absent. Everything—senses, mind, and feelings—was in a balanced state of concentration; and to separate these elements would have been to take from these children a perfectly natural way to discover what they had not known before. The unity of this triad is the essential ground for this kind of learning. (p. 41)

Our adolescent memories were scant with descriptions of flow and often revealed how emotions prevent pleasure in an activity. Gardening activities often bring states of flow to some people. In our adolescent memories of gardening, however, external reasons (mother made me) seemed to interfere with any expression or description of flow. In contrast, an experience of adolescent flow is evident in Rae's clarinet memory. She plays exercises over and over:

> Rae, Adolescent Air~Clarinet Memory: *[She] is entirely satisfied by her eventual control. . . . When she plays well, a deliberate evenness in her fingers on the keys, a bright wooden resonance, and the barely detectable thunk of the leather pads of the keys seating over the holes please her no end.*

While creativity is not evident in this memory, the pleasure is clear, and her motivation for meeting the challenge of playing difficult exercises keeps her returning to such experiences, all of which indicate flow.

Wonder

We believe that experiences of wonder facilitate the development of creative responses to the natural world. They expand possible interests and may contribute to feelings of freedom to pursue fascinating ideas. Among the memories were events that elicited moments of unexpected fascination, amazement, surprise or pleasure. In our early memories, we found, created, and invented opportunities in which we experienced wonder. As Lewis (1998) comments on childhood, "we lived by wonder, for by wondering we were able to multiply a growing consciousness of being alive" (p. 137). In Sue's early fire memory, wonder was evoked by a heat-driven carousel of Christmas angels.

> Sue, Child, Fire~Candle Carousel Memory: *Not much is clear about this amalgam memory except for the light, the movement, and the sound. The light is from the*

four candles that serve as the power source for the Christmas carousel of angels. As the windmill turns, the four angels with dangling golden rods lightly strike the two metal bells and make the most remarkably sweet sound. I remember sitting and watching the light and the movement, listening to the music it played, for hours.

We saw similar absorption in the wonder of the moment in memories from adolescence. Anna remembered her fascination with the behavior of magnesium. "*The material burned with such brilliance it appeared supernatural*" (magnesium memory). That fascination permeated her understanding of magnesium and its possibilities.

Wonder-filled experiences as adults also captured our attention during our analysis. Cele's account of the glacier memory reflects her absorption in a new experience. "*She had flown over the glaciers when she flew into Juneau, but she was not prepared to come face to face with the living, moving mountain of ice. It was a huge talking, living thing.*" The wonder of this experience transformed her perception of the glacier from something she saw from airplanes to something possessing human qualities. Her memory of grapes in the fall also describes wonder.

Cele, Adult, Earth~Grape Miracle Memory: *It was a beautiful fall day in West Virginia. The Young family had made a typical weekend trip to their farm up in the hills outside of Morgantown. On this particular day, the Youngs were joined by a group of colleagues and students, all of whom were wanting to enjoy the luxurious fall weekend. Their farm bordered the Cooper's Rock National Forest; even though it had been badly strip-mined and poorly restored (lots of Russian Olive trees to cover the carnage), it was truly lovely. On the south side of Pisgah Road was a single large tree and half-acre garden spot. This quickly gave way to a 300-foot drop, which, simultaneously with danger, afforded the visitors a spectacular view. The major part of the 40-acre farm was to the north of the road. It extended to the new freeway on the north, the national park on the west, and the cemetery on the east.*

Walks generally began on the old mining road, gradually curving uphill towards the northeast part of the farm. The road made a "Y" about one-half of a mile in and the left curve headed toward the national forest. Cele, Marie, and one of her students, Ollie, took the left road toward the forest, scaring off some of the deer loitering nearby. They decided to walk back on the west boundary of the property which quickly turned into another bluff, dropping approximately 100 feet.

After approximately a quarter of a mile, they all wanted to peer over the edge. Much to their surprise, they found what came to be known as the "grape miracle." The bluff was covered with native Concord grape vines which were brandishing their wares proudly. Instead of being green from the leaves, the bluff was purple from all of the ripe grapes. In astonishment, the group tasted the grapes, found them sweet, if seedy. They rushed back to tell the others of their find. Everyone gathered what containers we could muster and collected the wonderful fruit. Naturally, the wildlife

complained the whole time; the birds swooped and squawked and the squirrels chat-
tered. Even after we gathered all we could, both that day and the next, there was
more than enough left for everyone. Truly the legend of the "grape miracle" was
born.

A natural wonder, amazement, becomes the focus of the entire family group in
naming the experience a miracle. The lyrical quality of the grape miracle memory
reflects the strong feelings of importance, attachment, beauty, connectedness, and
the jewel-like aspect of the experience.

Perhaps the moment of recognition in Bell's earthquake memory gives us evidence for a connection between wonder and making new meaning or seeing the
events of the memory in another way.

Bell, Adult, Earth~Earthquake Memory: *It was unlike anything she had ever felt*
before. She likely would have just ignored it, but it continued to happen. . . . There
was that moment of puzzlement—"What is this?"—then that moment of recogni-
tion, but it was an odd sort of recognition because this was like nothing she had ever
felt before. . . . Without a word, they all seemed to get up at the same time and run
out of the building. You weren't supposed to run out of buildings, she thought, but it
seemed so instinctive. For all of them, it was the natural thing to do.

"Puzzlement" is a clue to the wonder in this experience. A new perspective cued an
instinctive response. Then, when she intellectually analyzed it, she criticized her
action.

Another adult experience of wonder is clear in Sue's recollection cued by air.

Sue, Adult, Air~Balloons Memory: *The whooshing sound . . . was unmistak-*
able. I threw back the covers and ran to the window. There was a cluster of approx-
imately 6 people in the field behind the house. They were artfully and quickly ready-
ing the balloon. The sound was the generator of the blower as it filled the balloon
with air. It took no longer than ten minutes and it was quiet again. The colorful orb
was rising slowly. . . . Although it was a slight disruption, as usual, I was not able to
return to sleep. The rush and wonder prevented it. I lay in bed and thought of the
height, the air, the wind, the jet stream, and fear, bravery, and luck.

The wonder of the balloons memory is powerful in expanding Sue's thinking
about hot air balloons. She describes new feelings, imaginative extensions of the
quality of the air, and romantic notions of the experience for the balloonists.

Bell, Adult, Water~San Francisco Bay Memory: *She was just emerging from the*
Caldecott tunnel through the mountain, and there it was, all of San Francisco Bay and
the Bay Bridge. It filled her with such an incredible feeling of elation and excitement.

She was there; she had done it. She was actually living in this beautiful, wonderful place, and it was a metaphor for the incredible freedom she felt and the incredible sense of possibility that lay before her. It was a feeling that would always come back to her every time she saw that view.

In this San Francisco Bay memory, Bell sees the Bay as the symbol of multiple creative possibilities. Opening the door to freedom, work, and self, Bell tied the wonder to a specific feeling.

As a culture, we have set up a false dichotomy between culture and nature (Lahar, 1993). In the West, the "progress" of Europeans in the Americas has been assessed largely through the development of Western culture here. Many of these focused accounts do not include the natural world, how it is affected by and how it affects culture, except in terms of how the natural world may provide useful material resources, for example, sod for building houses, timber for ships. The "taming" of nature including Native American cultures also has been a common motif. Our memories of interacting within nature are our best evidence of the intertwining of culture and nonhuman nature. At the same time, our narratives reveal to us how dualistic thinking begins to assert itself.

> To transform our relationship to the past by learning to understand the interactions and continuity of what has been divided into natural and social history, to establish a personal relation and place in it, is to develop roots—a metaphor that expresses grounding in both the organic world and social communities. This is riskier, more confusing, more exciting, and more transformative than adding on pieces to a purely social construction of history. It involves experiencing viscerally and intuitively, as well as rationally, the genesis of the human body and its organic and subjective evolution out of the oceans and savannahs, as well as through the social milieus of our grandmothers and grandfathers. We simultaneously arrive from the past and depart for the future in each encounter with history and with the decisions that we must make today. (Lahar, 1993, p. 114)

This chapter is meant to show that through memory-work, it is possible to reclaim what Egan (1997) calls the "intellectual tools" of childhood, the mythic, somatic, and the imaginative to reclaim the wonder so eloquently described by Lewis (1998) in his observations of young children. The cues we chose for memories (e.g., air, water) brought us beyond our causal narratives of self to memories we might have forgotten—memories collectively analyzed and theorized with the result that we have reclaimed wonder, exploration, creativity, and play. In this reclamation, we open up possibilities for knowing nature in different ways.

CHAPTER 7

Family Landscapes in Nature

One of our first breakthroughs in analysis occurred when Sue remembered that she did science experiments with her mother. This was a memory she had long forgotten, and since Sue's memories of the elements had primarily focused on her father, we wondered if the forgotten memory was evidence of how she had been socialized to think about the natural world, i.e., learning that nature is a place dominated by men. It was at this point that we decided to analyze the memories using family as a focus. We speculated that the relationships and related emotions in our childhood and adolescent memories would signify what values, hence what we learned from our families about the natural world. This would, in turn, reveal some of the ways in which our relationships to the natural world are socially shaped.

We first examine how our fathers influenced our relationship within the natural world and then how our mothers influenced those relationships. We also look at moments of resistance in our early memories where we ignored parental injunction and explored the natural world on our own. Siblings and friends play important roles in the development of our relationships within nature and we explore those as well. Finally, we examine the connections between the social and nonhuman facets of the natural in our chosen families as young adults.

Fathers' Influences

Our fathers are prominent in many of our early memories. Taken as a whole, our analysis of the memories suggests that the natural world took on particular kinds of values through their presence. They taught us important lessons there. In many of these memories, control is a prominent feature of our interactions with our fathers, and it takes on different meanings depending on the context in which it is exerted.

Sue's relationship with her father and the natural world is embedded in control issues. Her sandbox memory, from age 9, is illustrative.

Sue, Child, Earth~Sandbox Memory: Sue and her friend need the wetter sand to play successfully. Sue can solve the problem because the sprinkler is going on the lawn. She carefully, very carefully, piles sand on a plate and places it strategically on the grass within the sprinkler's shower path. Her father grunts disapproval at this solution. He demands that she take the plate back to the sandbox. He then moves the sprinkler closer to the sandbox and indicates that sand must remain in the sandbox. Sue then piles sand on the plate and places it on the wooden rim of the sandbox. This move enhances the chances of getting the sand wet because the edge of the sandbox is closer to the source of water, the sprinkler, than the box itself. Also, the sand is still "in the sandbox" by her definition. Her father did not like this strategy, kicked the plate back into the sandbox getting sand in both their faces. He then picked her up by the arm and drug her into the house, spanking her with every step along the way.

This memory is full of awareness of the natural world. Water, sand, sprinklers, and grass set the stage for this exchange with her father. She understands certain relationships between water and sand and can describe their interactions and recognizes the limits of her sandbox environment. At the same time, she disputes the rules of play associated with her sandbox. Her father generates and enforces those rules. He has control in this sandbox world; she does as well but only when she functions within his rules.

When we discussed Sue's memories, it appeared that her father had made the natural world an uncomfortable place for her. Sue turned away from the elements of the outside world and was driven inside both figuratively and in reality. She commented that, although for the rest of us, water was in a swimming pool, stream, or an ocean, for her it was in a teacup. Theorizing, we thought the outside natural world, as constituted by Sue's father, was defined by control. Sue was uncomfortable with this idea and when she rewrote this memory, control was specified in terms of rule-knowing. In life, she hates to play the "game" by the wrong rules, she wants to do it right, but with her father the rules were never predictable. He never taught Sue the rules of play, and, thus, she resisted and eventually withdrew. She literally went inside and moved her interests there as well. Inside, her mother taught her the rules and even allowed her to change them when she needed.

Cele's rewrite of her childhood stream dam memory shows control as well. This control, however, was of the element but is also evidenced in her father arranging an experiment for his young daughter.

Cele, Child, Water~Stream Dam Memory Rewrite: It was a Sunday spring morning. Cele had taken the 20-mile journey with her physician father over to the nearest hospital in the hope that they would have some play time after his morning rounds. She waited impatiently in the late 40s, early 50s black Ford they had. She

contented herself by watching the comings and goings of the other doctors but was ec-
static when she finally spotted her father emerging from the hospital doors.

Her father took her to a flat piece of land a few miles from the hospital. It was
covered with quarter-sized rocks, which squished when she walked on them. He took
her over to a small stream that ran through the piece of land. She wanted to play in
the water, but he suggested that they change the course of the stream. Puzzled, she
asked how they could do that. He showed her how to build a dam with sticks and
rocks. She helped her father build the dam and enjoyed playing in the clear, cold
water. When she tried to taste it, her father quickly scolded her and said the water
was bad, that it was mine runoff, and could make her very sick. Much too soon, her
father said he must leave but that they would come back next week and see if the
water changed its course. He asked her to be sure to note where the stream ran now.

A week or so later, they returned to the site. Much to Cele's surprise, the stream
had indeed changed its course.

In this memory, control of the element through diversion of the stream was central. Cele's father arranged the activity. She wanted to play in the water; he wanted
to engage her in diverting the stream. She wanted to taste the clear, cold water, and
he scolded her because it was mine runoff and poisonous.

In later discussions about our memories and the presence or absence of sensory
detail, Cele noted that in her first telling and in our early discussions, she glossed
over the sensory detail.

> There are elements of the water image there that are incredibly powerful for me that
> I should have gone on with but I do not. I got caught [up in the detail of] being
> scolded by Dad for tasting this mine water, and yet the important part was that it
> was cold, it was sweet, it was all of these things that were part of this fascinating
> stream, but I was so hung up on the relationship issues [it was not relevant].
> (4/11/94, p. 21)

For us our analysis here was quite powerful and illuminating. Cele's memory illustrates that our relationships within the natural world are embedded in and constituted
by our relationships with others, in this case our fathers. In addition, once we deconstruct the memory through memory-work, other aspects of the memory are reclaimed.
In this case, Cele reclaimed her sensuous enjoyment of the water.

In the rewrite of Anna's spring burn memory below, her father is afraid and angry,
and Anna responds first out of her own fear. She thought her father perceived her as
incompetent and a "*floozy.*" In contrast to Cele's memory though, the sensory detail
in Anna's memory is rich and not submerged by the feelings of fear of her father. As
a result of our discussions, Anna recognized that she accessed her memories through
her senses, and it was the vividness of her detail that brought the sensory aspect of
the memories to the attention of the group.

Anna, Child, Fire~Spring Burn Memory Rewrite: *Another spring ritual was beginning, and the fragrance of smoke lingered among the other spring odors. Smoke moved slowly because Anna's father picked a day without much wind to burn the grasses around the house. The fence lines couldn't be mowed, and the cows didn't seem to eat the long strands of blonde grasses along the perimeter of the homestead. The entire front acreage would be burned although the reasoning for it escaped Anna.*

After a winter of indoor play, school, and routine that was deadly predictable, the lure of playing with the fire as it licked its way under the swing was overpowering. Anna ran to the swing and jumped on the huge wood-piece that hung 30 feet from the top of the largest swing set in the world. Her father was competitive in his craftsmanship and probably thought he was providing athletic challenges to his daughters by building the crossbars of the A-frame way beyond an immediate reach.

Pushing the swing higher meant that Anna had an encounter with fire only once during each long arc as she neared the earth. The fire was nearing, and Dad started yelling above the crackling fire. "Hey, get down from there! Have you got no sense in your head? Don't you have the sense to know the floozy crap you are wearing to that school will burn like kerosene? Anna, wake up and get off that swing! That dancing can-can." Anna thought she heard an analogy of burning in hell and sin, but perhaps it was imagination or guilt. She knew better and completed her fun with a jump off the swing out of the range of danger. It was a pretty good jump, too.

As we have noted earlier, the sensuous relationship with the elements is prominent in our memories—Sue's cool wet sand, Cele's cool clear water, and Anna's vivid descriptions of the smells of smoke and the beauty of flame. Play is also present in these memories, particularly in Anna's. Even after the scolding from her father, Anna completes her fun with a *"pretty good jump"* off the swing *"out of the range of danger."* There is conflict between enjoyment in the natural world and structure introduced by fathers in the form of rules that were unclear for Sue, orchestration of events for Cele, and fear and anger for Anna. In our initial discussions of these memories, the conflicted emotional relationships defined and dominated the memories of the elements. However, as we reviewed the memories for similarities, contradictions, and missing pieces, our understandings reached beyond the conflicted emotions that originally defined them. This juncture highlighted the tension between therapy and memory-work (Schratz & Walker, 1995); we moved the analyses beyond the individual. The sensuous feel of the element moved to the foreground; play became important as well as excitement. Fatherly control moved to a much less prominent position. Through our analysis, we became conscious of how we constructed the natural world, and we also reclaimed our pleasure in it.

Fathers also taught us about the natural world. This circumstance is illustrated in Cele's stream dam memory and in Anna's memory below, cued by "tree." Here learning is the focus of the relationship, and the tree provides the background.

Anna, Child, Air~Willow Flute Memory: *It was a warm summer day on the west side of the hayfield. Anna, her sister Katy Beth, and their dad had walked out to the fence line between the Solomon sisters' farm and the Nelson family farm. The adult task was probably to check for necessary fence repairs, suspecting that this fence line in the trees and brush took a greater hit in the recent winter or spring storms. Goldie, the family golden retriever, was along and joined Anna and Katy Beth in inspecting the brush along the fence line. At one point, while impatiently waiting for Dad to tinker with the fence, Anna and Katy Beth asked to go home or get involved with the job. At this time, Dad took out the black, large jackknife from his right-hand jeans pocket and cut a piece of new shoot from a willow tree. After slicing here and cutting there, Dad slid the inside of the [shoot] out of its bark. One or two more cuts on the bark, then Dad blew on the hollow tube (now about six to eight inches long) and played melodiously. Of course, this amazed and delighted both girls, and another willow flute was made so the girls could each have one.*

In our discussions of this memory, what was prominent is Anna's surprise at her father doing something fun, being musical, and using his jackknife for something other than work. Anna learned what you could do with a willow branch, and she and her sister Katy Beth were amazed and delighted by this, but Anna also learned something new about her father. He could be playful and whimsical through the making of a toy for his girls. Fathers worked and played in the social world of farm life in Minnesota. Amid the routine of his work, he took time for flute-making for his girls. He taught Anna about the natural world and about enjoyment in that world.

Bell's early air memory presents a very different view of fathers and the natural world. In this memory, Bell's father sets the stage and does not interfere with the events being described.

Bell, Child, Air~Arm in Air Memory Rewrite: *She is sitting next to her father, but she is not with him. She is in her own world. It felt like late morning on a summer day where the heat and humidity were just beginning to be uncomfortable. Interstate 495, though, was a cool road that was laid through tall trees that stretched for miles on either side-lots of shade to drive through. Her window was open, and she was probably sticking her head in and out to let the force of the wind play with her hair. She was bored—the stretch on 495 always seemed the longest and went on for-ever. There was silence in the car, no radio, and Dad wasn't saying anything. Fo-cusing on the boredom always seemed so torturous to her and made the time stretch even further. She put her hand out the window and felt the wind play with her fingers. She opened her hand flat and the wind pushed it back. She could resist it and bring her hand forward to let the wind blow it back again. Then she put her arm out there with the palm of her hand down and forward. Her arm rode the wind like a surfboard, then she'd lift her hand and let the wind bring up her hand and forearm and sometimes let her arm be pushed back almost behind her shoulder. Then she would resist the wind and struggle to bring her hand and forearm back down so she could ride the wind again like a surfboard. She'd do this over and over again. She would get lost in the sensation of giving up control of her arm to the wind, letting it be whipped back and then struggling to bring her arm through the wind and lay it flat again so it could just ride through it. She was in another world, exploring the sensations of the moment, the sensation of her arm being controlled by something else. This world was removed from the inside of that car, removed from boredom and time. The novelty of the sensations, the novelty of that altered state only lasted so long. Soon enough she was back in the front seat and thinking about being bored.*

What is interesting about this memory is its contrast with the others. The vivid description of interactions with air, experimentation, and novelty are paramount. Her father has inadvertently created a space for exploration. This is another view of fathers' influences in the natural world. Bell is not engaged with her father, so she finds something to do with his tacit approval. Like many children, she finds something to entertain herself to fight boredom. The question, from a social constructivist perspective, is what is learned? What values are internalized? It is likely that Bell is learning a skill relevant to the value of occupying time and also learning about her range of freedom within the structure of her relationships to her father. He did not control this moment by insisting that she keep her arm inside the car.

In Anna's early breathing memory, her father, like Bell's, is present and is also instrumental to the events described albeit in a much more explicit way. Here again, a father provides the stage for exploration.

Anna, Child, Air~Breathing Memory: *One late summer afternoon, Anna, at a later preschool age, and her toddler sister, Katy Beth, found Dad in one of his usual resting places on the floor near the kitchen counter. He had been out in the fields during the day, making hay, and would come in for something to drink and a short nap. Dad could nap anywhere and one favorite afternoon spot was on the floor. He stretched out with his head on the bottom drawer of the cabinets that were handmade by his father, Grandpa Nelson, and identical to the cabinetry on the altar at the small country church. . . . Anna's Grandpa Nelson . . . died the year Anna was born. Katy Beth and Anna believed that their dad lay on the floor to play with them or nap with them. So, as the girls rested their heads on their dad's chest, they were astonished by the sound of his breath. In and out with rhythm and harmony to the heartbeat. It sounded like the air had to go a long, long way to or through Dad's body, and it would come out warm. "Was this the same air that went into Dad? Is the air going in the mouth and out the nose, or what? Does Katy Beth's air do the same thing and can I hear it if I listen to her chest?" The game of playing with the sound and touch of air continued.*

Anna's father willingly provided the creative space in this memory. She and her sister perceive him as being there for them, and they use him to play and explore air. In our discussion of this memory, Anna related her process of recalling it. She acknowledged the science in the memory and the idea that she was experimenting and asking questions of herself. She also experimented with Katy Beth, made her lie down and listened to her air going in and out.

In sum, then, our memories reveal multiple examples of the ways in which our fathers provided different opportunities for us to learn about nature. They control, they teach, they allow us to explore and play in nature, and, sometimes, as in Anna's memory, they become willing experimental subjects.

Mothers' Influences

The presence of fathers in our memories of the natural world brought us to ask why are they so present, and why are our mothers absent or peripheral to the memory? As children, some of us perceived that our fathers "worked" and mothers "did not" though as adults we know better. Some of our fathers were often scarce, and this made them memorable. They came home and into our days, and we took notice of interesting things with them. In farm families fathers are close at hand throughout the day. Most of our fathers were also more likely than our mothers to be outdoors. Encounters with them and the elements outdoors were more likely than with our mothers. However, in Bell's case, her mother was outside much more often than her father, and she appears, albeit peripherally, in most of her early memories. Do we not notice our mothers because they are usually there—in the background, reading a book, gardening, visiting with friends, cleaning, or seeing that something was on the table for supper? Theorizing about the invisibility of mothers, we are brought back to the "role that memories play in socialization and social control" (Schratz & Walker, 1995, p. 41) and the potential those same memories, remembered, explored, and transformed, have to interrupt our perceptions and discourse about family.

We spent time analyzing our mothers' peripheral roles, and additional memories and transformations emerged. Sue's mother was transformed from a tangential figure to a central figure in her relationship to the natural world. Collectively, we had theorized that Sue's relationship to the natural world was constituted by her father and the control he exerted when she was outside with him. When Sue re-wrote the sandbox memory, control was peeled away, and predictable rules emerged as a clearer articulation of her feelings. Her father taught her in a way that made it impossible to control whether or not she could succeed. When she went inside, however, her mother was there and taught her in a way that made it eminently possible to succeed. Her mother taught her the rules, and those rules al-lowed for control and choice. When Sue moved her interests inside, she chose to learn in the environment created by her mother. In this world, she did her own science with ant farms, gerbils, fish, and so forth. Inside learning with her mother was "natural, safe, easy, and fun." This transformation is powerfully illustrative of the social control that results from unexamined memories. This "inside science" was not remembered in the initial memories we generated, and in fact Sue judged her early memories as "not as good" as ours, because they were not "sciencey." It is worth pointing out that the forgetting of this memory may also have something to do with the social idea that as humans, we do not generally consider human-made social and physical environments as a part of nature, but, of course, they are quite natural.

This idea that our mothers are more influential than we had initially thought in our relationships to the natural world also emerges in Bell's memories. We initially noticed that Bell is, with few exceptions, solitary in her memories and very absorbed as she interacts with the elements. However, her adolescent growing rocks memory brings her mother's and Bell's characteristic absorption into sharper focus.

In spring and summer, her mother could work in the garden for ten to twelve hours a day. The family perceived this as her leaving and forcing them to fend for themselves around mealtimes. The conflict this created is what was initially re-membered, but during our discussion, we observed a parallel between the absorp-tion in thought that characterizes Bell's early memories and her mother's absorp-tion in her hobbies. In terms of values and socialization, we theorized that perhaps Bell learned the value of absorption and incorporated it into her own interactions with the natural world. This theory of absorption is useful for Bell in a biographical sense, but what is more important here is an idea relating to the larger question of socialization.

In our initial interpretation of these memories, we do not "see" that our mothers could have been influential. We have been taught not to see our mothers as poten-tial agents of socialization in the context of the natural world and science; that is the province of our fathers. Here again, when we rethink these memories, we begin to see the influence of our mothers.

Our work with Cele's memories reveals this same pattern. Initially, our discus-sions of Cele's early memories centered on the controlling influence of her physi-cian father. However, as we delved further and examined her memories within the context of all our memories, we noticed that Cele's memories are filled with social interactions. Social gatherings with family and friends provide the stage or background for her memories of the elements and her personal science. A fragment of her adolescent campfire memory is one example of this.

> Cele, Adolescent, Fire~Box Campfire Memory: *But the most wonderful thing, by far, was the way those knowledgeable women built the campfires. At our camp, there was something called the Order of the Arrow, which was composed predominately of counselors and advanced campers. It was an honor to be one of those. They were the ones who built the campfires.*

This memory led Cele to the observation that strong women were a characteristic of her childhood. She was surrounded by older and younger women, who were brought into her life through her mother. Her mother worked at the camp Cele attended and closely connected with powerful women in the small Kansas town where Cele was raised. In our discussion, we began to theorize that social relationships represent a value that Cele learned from her mother. The importance of relationship and its link to childhood experiences had been apparent to Cele, but linking the social to her early personal science and relationships to the natural world is a new understanding. Social relationship provides the backdrop for her interaction with the elements, and this opens up the idea that Cele's mother, like Sue's and Bell's, influences her relationship to science and the natural world.

Clever Disobedience

We were struck by the utility of Bachelard's (1964) phrase "clever disobedience" to describe a prominent feature of our memories. He notes that our first knowledge of fire comes out of "social interdiction" because the social prohibitions concerning fire are very powerful. "If the child brings his hand close to the fire his father raps him over the knuckles with a ruler. Fire, then, can strike without having to burn. Whether this fire be flame or heat, lamp or stove, the parents' vigilance is the same. Thus fire is initially the object of a general prohibition" (p. 11).

The influence of the social in our interactions with the elements and trees is circumscribed by adult perceptions of danger. Our early memories in particular show the interdiction of the social. They also show Bachelard's "clever disobedience." In order to obtain a personal knowledge of the phenomena, we escape our parents and engage with the element on our own, or we resist them and steal some time with the element before they are able to stop us. Even though there is danger, particularly with fire and water, there is also what Bachelard calls "reverie." We stare at the fire; we stare at a body of water in a state of reverie.

Through our families we learned about the rules for engaging with the elements. Some rules were meant to keep us safe from "dangerous elements," and some seemed arbitrary or meant to serve the interests of adults alone. Some of these rules were longstanding ones, and others were articulated in the moment. A longstanding rule in Bell's family was "don't play with matches." Sue's father articulated a rule of the moment when she was playing in the sandbox. She had placed a plate of sand on the grass, and he told her that sand "*belonged in the sandbox only.*" Sue resisted her father; she looked for a way around the rule.

What is interesting about these rules is that in some cases we defied them, and in others there is a complex mix of both resistance and compliance. In many of our early memories, underlying this resistance/compliance is a desire to obtain a personal or unmediated knowledge of the element. Anna's early fire memory is an example of resistance, but she ultimately had to comply. We noted earlier that Anna continued to "play" despite her father yelling at her to get down from there. She did get down, in her rewrite, but not before getting in a little more fun, with a "*pretty good jump, too.*"

We also provide evidence in our memories of our desire to experience the pleasure of the element for ourselves. Even though we know some rules will be broken, the lure of the element is our focus. For example, in Bell's club pool memory, she resists the rules and her fears in seeking the pleasure of water.

Bell, Child, Water~Club Pool Memory: *Bell is at the club pool with her mother for the afternoon. She's off standing at the deep end of the pool while her mother is kibbitzing with her friends in their reclining plastic woven lawn chairs. Her family belonged to this club for the two years that her father did financially well driving his one truck. The name of the club was either the Fairlawn Country Club or Fairlawn Pool Club.*

Bell is staring at the water, seemingly fascinated by the depth. All she can think about is jumping in to get a sense of the depth-get a sense of what it would be like to be totally immersed without touching the bottom of the pool. She thought it might be something like flying. She seems to almost envision her body gliding through the water. It felt like she was staring at that water for a very long time and tossing the idea back and forth in her head about jumping in. She felt some doubt, and the idea of jumping in scared her. She also thought about how much trouble she would get in for doing that. Only older kids were allowed in the water at the deep end. The lifeguard would get mad, and her mother would get even madder. She was on that edge of really wanting to jump in and feel what the deep water would be like, but the fear and trouble she would get in was holding her back. She dismissed the debate in her mind and jumped in. She went straight down like a knife through the water feeling the smooth movement and the water around her. She hit bottom with her feet and then she felt a hand on her neck and was pulled out of the water.

In their study of emotion and gender, Crawford et al. (1992) note that reflection makes an episode memorable, and the process of decision making in this memory is particularly explicit. The decision for Bell revolves around fear of the water, but wanting to experience the pleasure of the water. The debate also includes fear of her mother's anger along with the anger of the lifeguard. She resists both the fear of punishment and fear of the water and jumps in. In this next early memory, Bell's fear is focused on being caught by a neighbor who always tells her mother if the neighbor sees Bell doing something wrong.

Bell, Child, Fire~Matches Memory: *Its unclear whether it's winter or summer. The scene can be played in both seasons. Bell is on the other side of the Ford in between the tall manicured evergreen bushes lining the neighbor's lawn. She's got a book of matches, and this might be the first time she has ever lit them; what she wants to do is light every one. She's feeling nervous about being caught because she is not supposed to play with matches. She's also wondering whether Frances, the nosy, gossipy, and incessantly talky woman who lives next door, will catch her and tell her mother. Her mother always seemed to find out about all of the stuff Bell wasn't supposed to do. Bell lights the matches—one by one, although she's drawn to lighting the whole book all at one time. She's not sure what would happen if she did that— whether the whole thing would blow up and burn her hand and singe her eyes. There is a distinct image of the gray burned matches all lying in a pile and barely distinguishable from the gray driveway.*

In this fire memory, the breaking of the rules appears more intentional than in the pool memory in the sense that Bell likely spent some time planning the experience. She had to obtain matches and be certain others were not watching.

In both memories she fears getting caught, and in both of these memories, Bell is seeking the pleasure of the elements—fire and water. Both can be dangerous, and so there are rules for engaging with those elements—"don't jump in water over your head" and "don't play with matches." It is tempting to draw the obvious parallel to forbidden fruit. Our parents want to keep us safe; we want to experience pleasure and gain our own knowledge.

Cele's memory of eating vegetables in the garden (dirt potatoes) illustrates a further example of resistance. In her first version of this memory, quoted in chapter 2 "Making Sense," Cele describes not wanting to wash the dirt off the vegetables because the noise of the water pump would attract adult attention. She and her friend ate the unwashed vegetables, and Cele stated that the *"rich brown dirt simply added to their delicious flavor."* In Cele's rewrite of this memory, details have changed, but resistance to adult authority remains the central theme. In fact she now risks adults hearing the noise of the pump. Cele describes taking the vegetables to the pump and washing them off. The sensuous aspect of this memory remains, but the pleasure she describes in eating unwashed potatoes is lost. At the same time, in light of collective discussion, Cele extended her risk-taking stance in the rewritten memory.

Cele, Child, Earth~Dirt Potatoes Memory Rewrite: *The garden plot was south of the stable, but the pump for the hose was behind the stable, where the asparagus and rhubarb grew. Cele knew she should not touch these plants since she was warned they were poisonous. But she and Sarah knew how good the early radishes were. So the girls wandered toward the garden plot, keeping a close eye on the kitchen window, since they would be scolded for grazing the garden. Pretending to play in the rich dirt, the girls clandestinely pulled radishes and lettuces, just a few at a time. Then they would go to the pump, pull up the hard-to-pull handle and wash their booty and eat it quickly. The radishes were as good as might be imagined — small, plump, not bitter at all, just tart and crisp. The new carrots were small yet, but, oh my goodness, they were so sweet. Refreshed by these wonderful earth treats, the girls went over to the potato patch. Again, playing in the dirt, Cele and Sarah found some of the as-yet small new potatoes. With these small white treasures, the girls again wandered over to the pump, hiding the potatoes in their pockets. They washed the beautiful nuggets and delighted in the wonderful crisp taste.*

In all of these memories, we are breaking a rule in order to experience the pleasure of water, fire, and earth. We are asserting ourselves and asserting our freedom. In these memories, the natural world is a place away from the authority of our parents, a place where we can experience control and mastery. In our resistance, as Walker notes, "is the secret of joy" (1992, p. 279). Crawford et al.'s (1992) study of happiness offers some parallels to the pleasure and joy found in these memories. Their memory-work group found mastery in play memories that encompassed "freedom from constraint" (p. 85) and achievement.

A different kind of memory, Cele's cave memory, incorporates a forbidden adventure away from parents and fear stemming from challenge as opposed to fear of consequences.

Cele, Child, Air~Cave Memory: *Quickly they ducked into what was actually only a dugout on the side of a hill. Cele and Sarah pretended that they were being chased by "the bad guys," whoever they were. The cave was a place to hide. It was a bit surprising that they had managed to get to the cave, since it was on a rather steep hill above the local river, and neither set of parents allowed their children to wander far from their fishing spot.*

The change in air temperature was what Cele noticed most, since [the cave] did not go enough into the hill to change the light level much. Sarah, by contrast, was interested in the view from the cave. As Sarah called to her friend to look out, Cele replied that it scared her to look over the trees. Equally scary was the thought of getting back down from the cave, although Sarah had found it by coming down a slight incline.

Sarah said, "What a perfect place to scout the enemy! We can see everything from here." Both girls flopped down and began to explore the inside of the dugout. Graffiti of persons' names and dates covered the dugout, but there was no trash. Most of the names we recognized as friends and schoolmates. After a few minutes, we decided that we would bring supplies when we came back so we would be able to stay longer. With a pact to soon return, we left and rejoined our parents.

Seeking the pleasure of exploration as opposed to the pleasure of an element is what distinguishes this memory. The change in air temperature is what prompted the narrative, but exploration free from the constraints of adults is at its center. The fear in this memory is also different. It is related to scary heights rather than a fear of consequences for breaking rules. Crawford et al. (1992) contend that fear, a psychological construct that is traditionally defined unidimensionally, is gendered. In their work, they include fear memories from a men's group and find some interesting differences from their own memories of fear and other women's groups. The men's memories were justificatory in the sense that they explained their fear with precise measurements of how big something was, or how deep, or exactly what kind of spider they encountered. Peers were also a part of the memory in terms of "face-saving" (p. 92). Cele merely expresses her fear of heights; she does not justify it. She also recognizes that Sarah does not share her fear.

Themes of "clever disobedience" linked to seeking pleasure in the elements and trees are not nearly as prominent in our adolescent memories. There is resistance, but it is related to social interaction. Anna's motorcycle garden memory is about a forbidden tryst with a boy in the garden which serves as background rather than focus. It is the social aspect of nature that drew us in adolescence.

Anna, Adolescent, Earth~Motorcycle Garden Memory: *This is the first time a boy has come to the farm, without coming with his parents. Andrew drove up on his motorcycle. Anna's hands tremble; her stomach rolls; her head feels light. She better quit this planting (it is making her sick?). She thinks about walking over to the motorcycle. No, there's no blood in her knees, so she waits until Andrew gets to the garden fence. Awkwardly, they decide to walk down the logging trail through the mud. At the pond, Andrew throws aspirin "to keep the frogs healthy." He admitted that the aspirin was for the headache and stomach-ache created by coming to visit. Now relaxed, conversation continues with childlike games, laughter, and jokes.*

Anna defied her father. It was wrong to meet Andrew alone in the garden behind the garage, but she did it anyway. "I think the anxiety is here. My dad let me know all along that I am the bad girl." She had to deal with that guilt while considering the idea of "whether I can have this relationship [with Andrew]" (11/2/94, p. 15).

Other forms of resistance in our adolescent narratives are focused on chores. Bell hates the spring chore of helping her mother dig up the gardens. Sue hated cleaning the leaves from the basement steps and hated helping her parents plant ivy on a hillside. In adolescence, our parents engage us in trying to control the elements, and we resist the labor required in these never-ending tasks.

Family Differences

Throughout our group's work, our status as privileged white women tended to obscure our differences. We were so much alike, that we initially failed to notice our differences in class, ethnicity, religion, and sexual identity. They came to the fore when two of us expressed feeling invisible and marginalized within the group. Bell is an out lesbian, but we largely ignored the issue of sexual identity. Anna's roots in rural poverty distinguished her early experiences from ours, and yet we also overlooked the issue of class. Ultimately, these issues along with others were raised and discussed as we struggled to understand the larger context of each other's lives.

In our earliest memories, the differences that appeared to immediately impact our relationships to the natural world were the sizes of our respective families and the presence/absence of family members. Anna's early memories are filled with the members of her immediate and extended family as in her sauna memory. Here, she describes the sauna as "usually conducted in extended family units on Saturday." Grandma is there with her "long saggy breasts; the pails were left in the sauna (getting hot); children used only cold water; the water was piped up from the lake and smelled like swimming time." In spring burn, her father is there along with unnamed others, watching the dead grass burn in the spring. In breathing, she and her sister Katy Beth are listening to her father breathe while he sleeps. Her muck memory takes place at Spring Lake where the extended family gathered for vacation, and finally, in her flute memory, she is with her sister Katy Beth and marveling at the parenting skills of her father.

Cele also interacts with other family members in her earliest memories. She is with her grandparents in her memory of burning her feet on the hot grate of a floor furnace (chocolate milk memory) and with her father in her water and air memories (stream dam and underwater swim memories). A second air memory (cave) includes a friend as she notices the cool air of a cave on a hot summer day. In her early memories cued by tree and earth, she is also with friends—the red sugar maple, calf, and dirt potatoes memories. In Sue's sandbox memory, she is outside with her father and inside with her mother or, in her seeds memory, checking on the progress of germinating seeds in school with her classmates.

In contrast to the presence of family or friends in the above memories, Bell and Rae are alone in many of their memories. Rae's brothers are present in some, but many that are evocative for her are memories in which she is alone. This aloneness conveyed by Bell and Rae is a solitude that is experienced close to home and close to others. It may be that the urban/suburban environments for Bell and Rae led them to seek out solitary spaces. Alternatively, Bell and Rae might just enjoy spending time alone. In this solitude, Bell and Rae experienced more freedom and less constraint from adults in the natural world.

For some of us, siblings offer opportunities for learning about nature. They provide an apprenticeship for Rae in her paper balloon memory, and they provide teaching opportunities for Anna and her sisters as we see in the breathing memory and the magnesium memory. Siblings also provide opportunities for comparison. Rae compared the hole she had dug in the dirt to her brothers' dugout (digging memory). Boundaries with siblings are more permeable and less threatening than the relational boundaries with parents. In Sue's experience as a single child, she had fewer opportunities to explore boundaries and experienced different forms of emotional investment. Her memories revealed little exploration or close emotional interactions with others.

Our Chosen Families

Evidence of how we were socialized to think about nature continues to be prominent in our adult memories. Through analysis, we have learned that the boundaries provided by parents and siblings, seen in early memories, shift to ones we constructed as adults. The ways in which we participate in our own socialization as adults are different from those of our early childhood and adolescence. In adulthood we have more power, and the context that we find ourselves in, as well as construct, influences our understanding of our connections to nature.

In adulthood, our social contexts have different implications. Connections to nature necessarily entail family for some. For Cele, who married young and had children young as she began a career path, we found that this context influenced her understanding of nature. Her grape miracle memory, a gathering for her family, graduate students, partners, and children, illustrates this context. In examining Cele's socialization, she re-membered/reconceptualized nature. Simultaneously she had a child and, in graduate school, pursued research in infant development. She had been socialized, as most of us are, to see these two events as separate, i.e., her connection to nature was separate from her work. Nevertheless, her new baby became a part of her personal as well as her professional science. For Cele memory-work forged a link between her personal science and her professional science. Her perspective on nature has moved from a focus on the elements and nonhuman parts of the natural world to include a focus on infant-mother connections. For Cele today, mother-infant interactions are an important part of nature.

As a young adult, Bell was the only one among us who did not marry and start a family. She went off with a family of friends from college in pursuit of new spaces and new possibilities. Prior to memory-work, nature did not appear to be even remotely connected to these experiences; nature was the green world separate from self. For Bell, memory-work has interrupted this schism between memory and self; the natural world became the context for new spaces and possibilities. The analysis by the collective of the following memories provides the context for this shift in her thinking. We begin with her adult San Francisco Bay memory cued by water.

> Bell, Adult, Water~San Francisco Bay Memory: *She was just emerging from the Caldecott tunnel through the mountain and there it was, all of San Francisco Bay and the Bay Bridge. It filled her with such an incredible feeling of elation and excitement. She was there, she had done it. She was actually living in this beautiful, wonderful place, and it was a metaphor for the incredible freedom she felt and the incredible sense of possibility that lay before her. It was a feeling that would always come back to her every time she saw that view.*

Here, the sight of water and the bridge intimately connect her to this space of possibility, a geographic space that contains a new life away from her home, an emerging identity as a lesbian, in short, a new way of being in the world. Her candlelight vigil memory, cued by fire, portrays her sense that she will soon become connected with a new family and community here. The context of this memory is the assassinations of George Moscone and Harvey Milk in San Francisco.

> Bell, Adult, Fire~Candlelight Vigil Memory: *She wasn't out yet. But still there was an underlying connection to the demonstration that was held later that night. Thousands of gays and lesbians held a candlelight vigil downtown in front of city hall. She watched coverage of the event on TV. Thousands were marching with candles, and it was absolutely silent and completely dark except for the thousands of lights moving down the street. It was compelling and deeply moving for her.*

Our analysis of these memories, along with Bell's adult tree and earth memories, illuminates a shift from a view of the natural and social worlds as separate to one in which the natural and social are intertwined. The feelings of young adult freedom and possibility are unified with and magnified by the beauty of San Francisco Bay. Fire is the signal of connection to a community, a connection that is not yet obvious but certainly felt. Memory-work shifts the gaze so that what is once seen to be purely "social" is now understood as inseparable from its natural context.

Using the lens of family to analyze our memories of the natural world, we see how the Western culture's dominant discourse emerges through our families. We are shaped by this discourse, and we appropriate it as is clear from what we are taught to remember and taught not to remember. For example, we remembered that our fathers exerted direct control, taught us, and they allowed us to explore and play in the natural world. However, we largely forgot our mothers and other powerful women and how they taught us to think about and experience nature. We also resisted the dominant discourse and engaged in "clever disobedience" so that we could experience the natural world for ourselves. As we grew into adolescence, we appeared to "grow out of" our close connections with the natural world. Echoing the dominant developmental narrative, the social appeared to take precedence and perhaps in this growing division between the social and the natural, we were also mirroring the dominant mythology that humans can somehow divorce themselves from the natural. We seemingly carried this belief into adulthood, but memory-work shows us that the social and natural determine each other and are inextricably intertwined. We repeat, our understanding of how we have been socialized to think about nature has been altered.

In the next chapter, "The Power of Girls," we explore in depth an issue that underlies our "clever disobedience" as children. Our desire to resist and experience the elements on our own terms suggests that we took power in our relationships to the natural world. The discussion of our analysis now moves to this final theme.

CHAPTER 8

The Power of Girls

Our task is not to learn where to place power: it is how to develop power . . . genuine power can only be grown; it will slip away from every arbitrary hand that grasps it; for genuine power is not coercive control, but coactive control. Coercive power is the curse of the universe; coactive power, the enrichment and advancement of every human soul.

—Mary Parker Follett, 1924

Feminists writing on power since Follett continue to argue for alternative conceptions of power; ones that do not entail relationships of dominance or "power over." Miller (1982) rejects the idea that power must be defined as power over others. She offers an alternative definition of power as the "capacity to implement." Brunner and Schumaker (1998) differentiate "power over" from "power to." "Power over" usually involves social control and domination, whereas "power to" generally refers to joint production and social collaboration. Kreisberg (1992) pulls together much of the literature that challenges "power as dominance" and notes that this definition is incomplete.

> There is another dimension, or form, or experience of power that is distinctly different from pervasive conceptions. The ignored dimension is characterized by collaboration, sharing, and mutuality. We can call this alternative concept *power with* to distinguish it from *power over*. (p. 61)

Power is an essential feminist concern, and it emerged early in our analysis of memory. We recognized the sandbox memory from our first discussion of it as key to our understanding of girls in the natural world. Issues of control in that memory were discussed over and over during our years of analysis, as we talked about the pivotal role that power takes in shaping our experience in nature. In the course of our examination of transcripts and memories, original and rewritten, we found that the theoretical framework for power represented by Miller, and Brunner and Schumaker as well as Kreisberg is useful in interpreting our data.

In our memories, we discovered situations in which others exerted power over us (e.g., in the sandbox, WSI, sweaty palms memories) and in which others have tried to exert power over the elements (e.g., chat piles, stream dam memories). However, our memories also illustrate "power with," in Kreisberg's sense of the phrase. In sections that follow we use this distinction to interpret instances in which mastery attempts, giving control to others, lacking control, facilitating others, and occasionally taking control take place. Situations in which we see power with women and exchanges of power are analyzed as well. We found apprenticeship (Rogoff, 1990, 1995) a useful framework for thinking about exchanging power and offering power to others. We conclude the chapter with a discussion of the development of self and personal science seen through our analysis of power.

Power in the Memories

Power is interwoven throughout our memories in complex and contradictory ways. We appear powerful *with* nature in many of our memories (e.g., arm in air, fairy rock, San Francisco Bay memories) though we probably never spoke of it until we embarked on memory-work. We recognized powerful women who facilitated our activities in nature (e.g., box campfire, sauna, carrot water memories), and in our memories, we appeared to understand that nature has a power all its own (e.g., tornado, arm in air, candle carousel memories). The power that we felt in nature seemed to diminish when others began to intervene in our relationship to nature (e.g., sandbox, spring burn, stream dam memories).

The quality of power changes from early to later memories and appears to reflect socialization or fuller participation in the culture. As young girls we were powerfully connected to nature and gradually began to distance ourselves. Rae's willow tree memory in which she describes the area under the tree as *"a wonderful place, secret and green, smelling like dirt and green,"* or Cele's dirt potato memory, which expresses intense pleasure in eating early vegetables right out of the garden both show this early attachment to the elements. As we grew older, the elements were often secondary to other concerns. For instance, several of the adolescent memories were less about the elements than about issues of self-consciousness (Bell's matchless fire, Anna's motorcycle garden, Sue's bikini memories).

Early in our analysis we referred to power in the memories as agency following Bakan (1966). However, closer examination of his work suggested that his use of agency included a sense of separation that seemed foreign to our memories. Bakan notes, "agency manifests itself in the formation of separations; communion is the lack of separation. Agency manifests itself in isolation, alienation, and aloneness . . . in the urge to master . . ." (p. 15). Most of our memories are situated in social interactions with others, whether real or imagined. Indeed, shared experiences or goals and fostering the growth of another are common in the memories of several of the group (Rae's family planting, Anna's magnesium, Sue's cat prints memories). Although mastery is indeed an important part of power, we did not see the stark individualism that Bakan describes.

When evidence of control emerged in the analysis of our memories, it often appeared to stem from "power over," the kind of power that is usually noted in the dominant culture. We experienced a lack of control in relation to the dominance of others. There also appeared to be a lack of control, unrelated to dominance, in relation to the forces of nature.

One example of an adult memory that shows recognition of lack of control is Bell's recounting of a California earthquake.

> Bell, Adult, Earth~Earthquake Memory: *It was unlike anything she had ever felt before. She likely would have just ignored it, but it continued to happen. The earth was very distinctly rolling up under her feet, so much so that her legs and knees were pushed upward.*

Cele's adolescent recollection of the death of a tree that she expected would outlive her by many years, reflected this same awareness of a lack of control. "*Losing it was losing a part of me. . .* ," mirroring a lack of control she felt over the timing of her own mortality.

Interaction with the elements most often involved other persons in our memories. When those interactions were with parents, they sometimes imposed control and/or rules. In instances in which control was taken from a child by a parent, emotional distress often resulted. Sue describes being held up in the air by her father:

> Sue, Child, Air~Up in Air Memory: *The child is being held up in the air by a hairy arm and wide-spread hand. Her father is lying or sitting on the floor holding her above him. No other means of support exists. She can hear laughter, that of her father and her mother. She sees nothing but the arm below her and the body. Her hands clutch at the arm for additional protection and security from the height. The experience lasts an eternity, probably one minute in length. She wants down desperately, but her father and mother act as if she should enjoy the experience.*

When we discussed this memory, we recognized the importance that control played even for a very young child. In the rewrite of this memory, Sue recognizes that she was powerless and has no chance to take power. She also hears the strong message that she should find this experience pleasurable.

Sue, Child, Air~Up in Air Memory Rewrite: *She is small. She is wary. She is being lifted above the terra firma and is out of control. Someone else, her father, is deciding what is happening above the ground. Only his strong arm supports her as he lies on the ground but keeps her from that same ground. Her father laughs and cajoles her. Her mother laughs and cajoles at the periphery of this scene. But, she is terrified and grasps at the supporting arm. She wants to come back to the ground but is in no way able to control that occurrence. She is also confused by the inner message of fear and alarm and the outer atmosphere of fun and frivolity. When finally she is returned to the ground, she reclaims her self and then worries about her inability to contend with the conflicting messages from self and others.*

In this instance, it is difficult to experience pleasure without control or power. Sue's father and her mother "*at the periphery*" expect this activity should be fun. For her part, Sue is confused by the conflict between her fear and the fun her mother and father are having and expect their daughter to have. Sue resists and wants to end the activity, but she also remembers feeling guilt for not complying with a social expectation and guilt that her feelings were at odds with her father's. At the end of this rewritten narrative, Sue describes regaining power in terms of "*reclaiming her self*" when her father returns her to the ground. However the power seems short lived. Instead of feeling anger at this imposition, she feels anxious and confused for not feeling what others feel. The expectation that she should feel what others feel is an early lesson in gender training (Brown & Gilligan, 1992; Gilligan, 1982).

Another childhood memory illustrates how reactions to being controlled by arbitrary rules were sometimes met with clever attempts at resistance. Sue attempts in ingenious ways to cope with her own needs for sandbox play and her father's insistence that the sand be kept in the sandbox.

Sue, Child, Earth~Sandbox Memory: *She carefully, very carefully, piles sand on a plate and places it strategically on the grass within the sprinkler's shower path. Her father grunts disapproval at this solution. . . . He then moves the sprinkler closer to the sandbox and indicates that the sand must remain in the sandbox. Sue then piles sand on the plate and places it on the wooden rim of the sandbox. . . . Her father did not like this strategy, kicked the plate back into the sandbox.*

Although Sue's solutions to her father's rules ultimately led to punishment, she did try to construct the situation so that she could succeed. Crawford and her colleagues (1992) described similar memories from their collective as reflecting the powerlessness of the child to change the rules. Sue noted in discussion that in asserting herself she protested the injustice and incompetent use of the rules.

Power of Women in Relationships

Authors who have worked to develop alternative conceptions of power have noted that many women view and use power differently from most men (Brunner & Schumaker, 1998; Helgesen, 1990; Kreisberg, 1992; Rosener, 1990; Schmuck, 1999). Miller (1986) believes that women start from a position of being dominated and must learn to resist others' attempts to control and limit them. Kanter (1977) believes that power should be conceptualized as the ability to get things done based upon relationships and participation. This idea is similar to Miller's (1982) conception of power as the "capacity to implement." Arendt (1972) claims that "power corresponds to the human ability not just to act but to act in concert. Power is never the property of an individual; it belongs to a group and remains in existence only so long as the group keeps together" (p. 143).

According to Brown (1998), girls compromise their power and what they know about themselves and the world through experience; they adapt to cultural conventions. Direct evidence of this lies in Sue's eventual move from outdoors to indoors after the experience with her father in the sandbox memory. She left the ambiguous rules of her father's realm of nature outside to go to her mother's clearly defined indoor realm of nature.

White middle-class girls during the early school-age period move away from the evidence of their senses in Brown's view. We came to believe that this statement was often true for us. Although issues of control, or lack of control, seemed to dominate Sue's sandbox memory, when we analyzed and discussed the memory, we recognized that Sue was paying close attention to her senses as well. This brought the sensuous to the foreground, and Sue's personal relationship with the element became obvious. Through memory-work, we stripped the dominant control issues from the experience, and Sue's personal science emerged. Here we saw how socialization could cloak relationships with nature as it did initially in covering the sensuous in Rae's red boots memory and in Cele's stream dam memory.

In further discussions of the sandbox memory, Sue noted that when she was inside her house, her mother structured clear rules within which she could succeed. Follett (1924) likely would see in Sue's father's behavior the coercive component of control. Coactive power, in contrast, is the ability to make things happen with others as is illustrated by Sue's mother's approach. When asked to rewrite this memory, Sue focused on control in terms of knowing the rules as we indicated in chapter 7, "Family Landscapes in Nature." In the ambiguous outdoor world, Sue felt she had little power; indoors, where the rules were clear, she had a sense of greater control.

We were impressed by the power of relationships with women and the powerful women in our lives. Cele's friend Sarah figured prominently in her early memories (cave and dirt potatoes memories). She and Sarah would not have ventured into the cave without each other's urging. Together they also sampled the purloined potatoes from the garden. As adolescents, Anna and Fran were in charge of "*the live tree and stream scene*" for the Junior Prom.

Anna, Adolescent, Tree~Junior Prom Memory: *This year the big event was planning for and going to the Junior Prom. As members of the planning committee, Anna and Fran were responsible for decorations. The theme had something to do with the northern outdoors and a live tree and stream scene was to be created in the corner of the cafeteria for the big night. Anna and Fran spent Saturday on the back roads of the reservation looking for the right size, shape, and quality pine trees for the prom scene. Conversation was loud while they slowed their motorcycles at each pine stand to search for the spot where the work would be done.*

Anna and Fran were the only girls in their grade with motorcycles; together, they took off into the countryside to survey various stands of trees. After choosing their favorites, they felled the trees by themselves and coordinated carrying them back to the prom site. Their bikes gave them freedom to roam the countryside, and they accomplished their tasks by joining forces.

In some situations, the power that emerged through friendship took a different form. A variation on joining forces is seen in Bell's dunes ranger memory, which documents the power of friends in eluding authority to achieve their goal of sleeping on the beach.

Bell, Adult, Air~Dunes Ranger Memory: *She was with Nan and they had Katy [Bell's dog] with them out at Point Reyes National Seashore. It was very late in the day, and they were walking the beach and the wind was ripping at them. You had to point your body into it to make any distance walking. It was a grey day; the sky was totally socked in. They sought relief from the wind and went up into the sand dunes. It was almost like shutting a door; the wind had stopped and they could hear each other speak. I think it was Nan who suggested that they just stay the night at the beach. She always carried sleeping bags in her car, in her old blue and very beat-up Volvo. It was illegal to camp on the beach, so we waited until dark to go back to the car and get the bags. . . . Something woke Bell up in the morning; it was still grey and the wind was still blowing. She poked her head just above the top of the dune and saw a green-uniformed ranger. Katy was alert and starting a low growl. She rolled on top of her and held Katy's mouth shut trying to shush her. If they were found, they would have been fined and kicked off the beach. The ranger kept walking and never saw them.*

Strong adult women, not only mothers, were important to us. In Rae's adolescent carrot water memory, this is clear, "*She really wants this to work both to see it herself and because she wants it to be right for Miss Schwartz, the biology teacher.*" In our analysis of memories and subsequent discussions, we found adult women to be role models and rule makers for us in adolescence. As Cele said in her box campfire memory, "*the most wonderful thing, by far, was the way those knowledgeable women built the campfires.*"

The importance of the support of other women is shown in Bell's club pool memory. Bell's mother not only went to the pool for her children's entertainment; she loved getting together with her friends. In Bell's childhood memory, she reports that her mother was "*kibbitzing with her friends in their plastic woven lawn chairs while Bell's off standing at the deep end of the pool.*" In further analysis of gatherings of women in our memories, we were reminded of women's groups from earlier in the century such as Home Extension Clubs or even the Women's Christian Temperance Union. Such groups provided support for women's interests by providing them with a socially acceptable outlet for their intellectual and political growth.

In our discussion of Anna's memories, we were struck by the parallel between a local Native American women's club and her mother's rural club in Minnesota, both lasting over 50 years for the women involved. Cele noted that in a recent obituary from her small Oklahoma town, the first detail reported was that the woman had been a member of the home extension club for 74 of her 84 years. The women in these clubs were like family to each other; they supported each other, and they knew each other's "business" as well. Through the clubs these women facilitated one another's growth and often provided services to the community.

We discussed the parallels between these women's clubs and our own collective. We acknowledged the power that each of us gained from our work in the collective. We connected with each other, catching up on one another's lives at the beginning of our work sessions, and occasionally recent events in our lives became the central focus of a meeting. The importance of the group, both personally and professionally, was clear in our commitment to frequent meetings and work over such an extended period of time (1994–2001).

Exchanges of Power

As women with families, all of us reported making decisions with family interests in mind. In telling descriptions of our careers in relation to the development of a family or a spouse, three of us noted that careers were placed in a secondary role. Coupling often seemed to reduce one's own power. For instance, Anna notes, "The whole idea that I always have to take second. . . . In our family, it was whoever made the most money . . . made the decision [about career moves]. It's the same thing" (10/27/95, p. 3).

We discussed turn-taking in career decisions with partners, which involves acquiescing to another's control. Some decisions were influenced by the social context of the times. Rae, speaking of the 1960s, said, "And I really in those days believed that the decision about the job and the move was his to make. 'If it's the best job, Eric, if you think it's the best job, you go'" (10/27/95, p. 9).

These decisions often involved considerable career sacrifice. Rae, in two of her adult memories describes her moves with her family. In her recounting of her flying memory, Rae says,

> Rae, Adult, Air~Flying Memory: *She has told Eric that the decision about the job and move are his to make; whatever is best for his career is ok with her. But she is not at all sure she really wants to live in Mississippi, just up the road from where three civil rights workers were shot.*

Poignantly, in her adult beds and trees memory, Rae is "*deciding where the furniture is to go in the 8th place she has lived in during the 10 years she has been married.*"

Virtually all our memories reflect relationships with others. However, Bell's memory of a move to the West Coast is a singular example of autonomy.

> Bell, Adult, Water~San Francisco Bay Memory: *It filled her with such an incredible feeling of elation and excitement. She was there; she had done it. She was actually living in this beautiful, wonderful place, and it was a metaphor for the incredible freedom she felt and the incredible sense of possibility that lay before her. It was a feeling that would always come back to her every time she saw that view.*

As discussed in the section on mastery to follow, most of our memories demonstrate strong ties to the lives of others. Bell's memory surprised us. We were impressed by her determination and ability to make autonomous choices as a young adult. The rest of us felt that we could not have made choices like that without considering the lives of others. Still, Bell did not make her choices entirely alone. She moved to San Francisco with five friends.

Apprenticeship within Families

The exchange of power through apprenticeship provides another venue to explore how we are socialized in relationship to power. Parents often structured early experiences by restricting the choices we made as children. In the underwater swim memory, Cele's father was playing with a group of children in a swimming pool. Cele, trying to swim across the width of the pool underwater, was excited by her eventual success. With only brief acknowledgment of her accomplishment, her father urged her to swim the length of the pool instead of letting her swim the width again. The closely directed activity, an example of Rogoff's guided participation, yielded little opportunity for active structuring or power for Cele.

In a series of three early fire memories, Rae experienced a bit more power with her older brothers who apprenticed her in her interactions with the element. Siblings may sometimes share more power than do parents. In the first memory, she is 6.

Rae, Child, Fire~Candle Flame Memory: *She sits at the supper table watching her brothers pass their fingers through the candle flame. She really wants to do this, and so she does, delighted and surprised at not really feeling the flame as her finger passes through it.*

Then she is 8.

Rae, Child, Fire~Cigarette Memory: *She is watching her brother Sam getting ready to light a cigarette. They stand in the basement, which her brothers have recently fixed up. Beer kegs, tables, and chairs are painted on the walls. She and her brother stand near an old white enameled gas stove. Her brother strikes a match without closing the cover of the matchbook, and the whole thing bursts into flame, burning his hand. She is amazed and frightened at what could happen striking a match.*

Then she is 9.

Rae, Child, Fire~Paper Balloon Memory: *She stands in the empty lot next to their house. It is fall, the leaves dry and brown, rustling underfoot when she and her brother Fred walk to this spot. He lights a tiny fire of leaves under a balloon he has made by folding a newspaper into a ball with an opening in the bottom. He explains that the air will heat and raise the balloon into the air. It does not happen. The paper balloon does not rise and Rae feels sorry for him.*

The apprenticeship in these memories is both intentional and unintentional. In the first two, Rae learns by watching her brothers as they interact with fire. In the third, her brother Fred wants to show her what heat can do, but the demonstration fails. In an early air memory, Rae describes watching a thunderstorm from the shelter of a garage with Fred. Air triggered this memory of the smell of rain in the air, which serves as background for learning about electrical storms.

Rae, Child, Air~Thunderstorm Memory: *They watch the thunderstorm together. She loves the dark clouds, the smell of the air, the lightning and thunder, the feel of rain, and Fred's company. They count the seconds between lightning and thunder, judging how far away the storm is. She knows she needs to practice this so she can tell how long it's safe to stay out in the storms.*

When we analyze the values conveyed and learned in these memories, they center on the idea that the natural world is something from which you can learn. This is similar to the value that Cele's father tried to convey in the stream dam memory, but the element of control he imposed complicated the learning and created conflict. In Rae's series of memories, this conflict is not present. The emotions around her relationships with her brothers are a part of the memories; they provide a supportive, yet emotional backdrop—comfort with Fred, feeling sorry for Fred, amazement and fear around matches and fire. These emotions are integrated with the focal objects of the memory and in that integration, values associated with fire emerge. Rae learned about fire through an exchange of power. Her brothers apprenticed their younger sister as she gained some competence with fire.

Particularly useful for us in our analysis has been the framework provided by Rogoff and her colleagues (1990, 1993, 1995) to describe the multiple planes of apprenticeship and how children learn to use the tools of a culture. Rogoff (1995) sees sociocultural activity as happening on three planes: guided participation, apprenticeship, and participatory appropriation. All were salient in our memories.

In our early memories, some of our activities clearly were structured as in Cele's stream memory, Sue's sandbox memory, and Anna's sauna and breathing memories. Rogoff and her colleagues emphasize the importance of observation in guided participation. Observation played a key role in the memories whether guided, when Cele observed the changes in the stream as instructed by her father, or unguided, when Anna observed the changes in the flow of air of her father's breathing.

Adolescence brought additional responsibility and an active role in structuring activities, which Rogoff describes as apprenticeship and participatory appropriation. Often, that role was part of an opportunity to demonstrate the skills of the culture as in Bell's WSI and Rae's Bunsen burner memories. In adulthood, we provided opportunities for apprenticeship by structuring activities or sharing understanding. Rae's family planting and Bell's role as a teacher in her earthquake memory both illustrate these. We gained power as we directly took on the role of transmitter of the culture and our values. We also then could offer an opportunity for power to the next generation.

Power as Competence or Mastery

We detected a sense of power in our early memories that is akin to Miller's (1982) "capacity to implement" in relation to the element. Kreisberg (1992) explores the etymology of power and notes that "the word derives from the Latin *posse*, which means to be able" (p. 56). As he points out, that Latin root is reflected in the definition for power as "ability to do or effect something or anything, or to act upon a person or thing" (*Oxford English Dictionary*, 1989, volume 12, p. 259). We found delight in our ability to do or effect something in relation to the elements. In Cele's stream dam memory, she was excited that she had been able to change the course of a stream. Similarly, Anna was very interested in the understanding of the element she acquired as she listened to her father in her breathing memory (Anna, child, air). The pleasure, even joy, associated with the experience of implementation or engagement with the element seems central to these memories.

Early exploration and perhaps recognition of the power of the element was evident in Bell's early arm in air memory. In this excerpt, Bell plays with the air and explores the element with little parental facilitation. Her personal science, as discussed in chapter 4, "Making Sense," is at work:

> Bell, Child, Air~Arm in Air Memory: *Her fingers are hanging on the bottom edge of the window, and she starts to lift them and she feels the force of the air. If they were more flexible they [would] bend backwards. Then she moves her hand into the wind force, and she can allow it to be picked up and tossed by the air. Soon she's got her whole arm out there, and she gets totally lost in allowing her limb to [be] played by the air.*

In the later memories of times when most of us had some training in our professions, we found surprisingly few descriptions of feelings of power or competence, especially like those found in the early childhood memories. Rae's adult quarry pond memory in which she was practicing as part of her professional training is one of the few examples of feeling power or competence. In that memory, she was the person who structured the experience, and she described her success in the endeavor. By contrast, in several adolescent memories, our feelings of competence were dependent on demonstrations or judgments by others as we commented in the efficacy discussion in chapter 5, "Metaphor."

> Bell, Adolescent, Fire~Matchless Fire Memory: *It was possible she thought to start a matchless fire—an amazing feat in those damp woods, and it would be the first time she had done it. Though the accomplishment wouldn't be as good as starting with two sticks or a flint and a stick, something that absolutely amazed her. She started to shape a tepee over the ember, leaving an opening so she could get tinder in there to catch on the ember . . . a matchless fire she told everyone as they gradually arose. She staked out new territory in the campfire competition.*

A sense of our own mastery emerged in our adolescent recollections; we were conscious of our own skills and abilities, our strengths as well as our weaknesses. Anna (magnesium memory) and Cele (in discussion of her watermelon memory) described their academic abilities. Rae knew when she played the clarinet well (clarinet memory), and Bell predicted her weakness in her lifesaving skills (WSI memory). We seemed to be able to differentiate realms of our performance and thought; we could measure our strengths and gauge our successes.

Harter and her colleagues (1997) have described the development of multiple role-related selves during adolescence as well as the ability to resolve contradictions in these self-perceptions during adolescence and adulthood. The multiple selves of a 15-year-old described by Harter are not unlike those reflected in our own memories. For example, Anna dealt with the incongruity of her seeming inability to waterski and her excellence in high school science classes. She was both incompetent and competent, depending upon the "self" that she was describing. While our memories are discrete narratives of events in our lives, the evidence from them suggests that we as adolescents and adults, especially, differentiated the realms of our successes and failures.

A model of the construction of self that integrates autonomy and connectedness has been provided by Harter (1999). She noted that earlier Western theories of self-reflected values such as autonomy, independence, and self-focus, which implied developing relatively impenetrable boundaries between the self and others. Recent versions of this position, however, have emphasized that healthy adaptation requires an integration of self with others to assure self-esteem, authenticity, and good mental health. Hoagland (1988) described this intertwining of the social with autonomy as autokoenony: "A self which is neither autonomous nor dissolved: a self in community who is one among many" (p. 12). We noted that almost all of the memories displayed autokoenony in our blending of our lives with others. We seldom described autonomy outside contexts that included others' lives. Our memories reflected the intertwining of our lives with others.

In contrast to positive experiences of mastery in adolescence mentioned above, we also reported experiences of failure as judged by others. Bell (WSI memory) described the lifesaving test in which she was paired with the "lead weight" of the group.

Bell, Adolescent, Water~WSI Memory: *Her partner jumped into the pool and started thrashing. She dove in and made the correct approach. Got behind him and grasped him. She controlled him, but she couldn't get him high enough in the water. Water was in his face and up his nose. It was too sloppy and took too long to get him to the edge of the pool. When she did get him to the edge and got him in position to be lifted out—she couldn't do it. She could not lift him out of the water.*

When they were all tested, he [the instructor] left them there in the cold dampness around the pool while he went into his office to tally up the marks on his clipboard. It didn't seem like a long period of time when he emerged. He called them up to him, one by one, and told them whether they passed or failed. Lots of smiling faces. Then it was her turn—she had failed and to her it was her first moment of failure.

Anna's adolescent inability to successfully negotiate waterskiing led to a similar emotional response.

Anna, Adolescent, Water~Waterskiing Memory: *The trials were painful. How could water be so firm, so hard? Even the spray from the motor seemed to slap Anna across the cheek. Why was she sinking so readily when Bob literally "bobbed"? Anna suspected it [had] something to do with the weight of the water and her self-perceived weight problem.*

Learning a skill includes elements of practice such as the preparation for a singular event like Bell's campfire and WSI test (matchless fire and WSI memories). The opportunities to practice, to learn how to do specific skills, and to demonstrate them were important in our developing relationship to nature. It was clear that Bell engaged in practice and apprenticeship in campfire building. Similarly, Rae's description of lighting a Bunsen burner in chemistry class implies practice.

Rae, Adolescent, Fire~Bunsen Burner Memory: *She wonders whether she will be able to strike a spark fast enough after opening the gas valve to light the stream of gas at the burner before it accumulates and explodes in her face. She scratches [the sparker] fast as she can, lights the gas, and relaxes enough to breathe again .*

We observed that practicing a skill sometimes enables more power in relation to an element and hence more competence in the experience.

In the chemistry memory in which Anna observed her chemistry instructor light a magnesium strip, not only did apprenticeship occur, but also Anna's demonstration of the skill to interested others. She valued this skill as well as the magnesium as evidenced by where she kept it.

Anna, Adolescent, Fire~Magnesium Memory: *Anna managed to get an envelope of magnesium and took it home to demonstrate the wonder of the burning of this material to her sisters and others who would watch. Anna kept the magnesium in her jewelry box.*

Not only did she learn to ignite the magnesium, her sisters marveled at the same feat that her instructor had accomplished. Anna mirrored the mastery or power of her instructor.

In our adult memories, other aspects of mastery emerged. Some of us experienced mastery when we were nurturing our children. Rae described fostering the development of her children, a kind of apprenticeship, in her family planting memory. She managed to get the infant to sleep so that her other two preschoolers could help their father plant flowers in the springtime. She enjoyed "*their mutual pleasure in the task and her own sense of satisfaction in facilitating.*" Rae explained that she arranged the task "so that these kids can help him do that. I managed that. They had a good time. . . . And I did it" (10/27/95, p. 6).

Rae also described a sense of mastery in coordinating the mundane tasks of everyday family life. Rae had three children in less than three years and only one high chair:

> So, Grace was the youngest, so she would be in the high chair, and the other two would be at the table. You know, they knock things over all the time. So I went and got some . . . cookie sheets, where the sides go up and there's a sealed rim around them. . . . Then I got some suction cups with bolts and I drilled holes in those [cookie sheets] and then put the suction cups on the bottom and they go on the table and we put the kids' food [in their plates and glasses] on top of that. So here are these two little girls up here on their stools, and they could not make a mess by spilling their food on the table or the floor. That was the zenith of my cleverness. Don't you think that was pretty clever of me? I told you I was clever. (10/27/95, pp. 6–7)

Similarly, in her flying memory, Rae described traveling with her infant. After boarding, she nursed the baby and the child fell asleep; Rae was able to rest the breakfast tray directly on the three-month-old infant, eat her own breakfast, and enjoy the flight. Cleverness and creative problem solving are evident not only in early childhood but also in some of the adult memories. Some aspects of mastery in adulthood reflected our willingness to structure the environment for others' apprenticeship or for our own satisfaction. We used power and engaged in the cycle of providing opportunities for others to exercise power. When we analyzed these memories, we were impressed by our own abilities and power.

Conclusions

Power as domination is pervasive in the world. It is a facet of socialization, and thus it affects our relationship to the natural world. This kind of power is evident in our memories, but it appears to be far less prominent than power with others and, we might add, power with nature. Certainly, we feel power with the elements in such memories as the fairy rock, San Francisco Bay, willow tree, and quarry pond memories. We also feel the power of nature in the earthquake, tornado, and chat piles memories, and we experience power through and with other women in the junior prom, box campfire, cave, carrot water, and dunes ranger memories. When we experience power as domination as in the sandbox memory or for safety reasons in the spring burn and stream dam memories, we seem to lose or forget our connection to the element, but memory-work shows that this loss can be recovered. In fact, Starhawk (1987) observes that we experience these connections or what she calls "power-from-within" frequently.

Although power-over rules the systems we live in, power-from-within sustains our lives. We can feel that power in acts of creation and connection, in planting, building, writing, cleaning, healing, soothing, playing, singing, making love. (as quoted by Kreisberg, 1992, p. 68)

Through memory-work we found that we experienced power in the personal science of our early memories and that the quality of this power changed in our adult memories. We moved away from our power as young girls and our personal science. Through our analysis, we recognized that the personal science that we constructed as girls was something we could control, although sometimes it was only within the limits that adults structured for us. As young girls we had power to construct a personal science we "owned." As we grew older, we lost some of our power in nature and forgot our personal science. We were taught in school a science that was almost exclusively structured by others; we were apprenticed and became skilled in the methods of this science, the tools of our culture. In some ways, though, traditional science is no longer our science. We have begun to acknowledge the interrelationships of power and apprenticeship. We are more keenly aware of both our power and the influence of the tools of our culture, particularly as we think about our professional lives.

Power and apprenticeship are significant in the construction of both our personal science and professional science. Although some of the linkages are less direct than we originally envisioned, a number of interesting considerations emerged in our analysis. For some of us, control of the elements or the environment was crucial in the construction of our personal science. This control was seldom autonomous and often exerted within the context of our families. For Sue, that meant that her science, whether personal or professional, was moved indoors where rules were clear. For the rest of us, our apprenticeship and mastery in the outdoor environment also occurred within the context of our families and friends. Our successes and failures were not only our own; they were linked with others as well.

Our adolescent memories provided examples of stages for us to show our competence. The practice and performance of mastery were important in these memories though only a few of these memories narrate demonstrations of our competence as scientists. Likewise, only a few of our adult memories show examples of our competence in our professions. The power we felt in our personal science appeared to wane in our adolescent and adult memories. However, some indirect linkages from our personal science as children to our professional science in adulthood emerged as we talked about our relationships to the elements and to science. We have blended the fascination from our childhood science into our adult science. We were apprenticed in the tools of the scientist, but we use them in ways that are linked to our personal science. For instance, some strong interests and connections (e.g., Bell's with thinking, Cele's with social interactions, Rae's with water, Anna's with teaching, and Sue's focus on rules) have direct links to our current scientific interests. We had forgotten our power as girls, and through memory-work we recovered that power in our lives and professions as adult women.

Memory-work has illuminated the power not only of the culture but also of the individual in constructing an individual's science. However, the complexities of both power and apprenticeship are important considerations in understanding the development of that science. Power implies more than "power over," it includes understanding the rules of power, the emotional aspects of power, the recognition of power in the elements, and the giving of power to others. It includes demonstrations of mastery for oneself as well as for others. In the analysis of our memories, apprenticeship emerged as more complex than Rogoff's (1990, 1993, 1995) description of guided participation, apprenticeship, and participatory appropriation. It was linked with the ability to exercise power, mastery, and competence as well as the capacity to continue the cultural cycle by giving power to others. As we conclude in chapter 9, "Interruptions," the analysis of our power and apprenticeship established a bridge between our personal science as girls and our professional science as women.

CHAPTER 9

Interruptions

By using discrete memories of moments in time, collectively joining subject with object, and linking personal experience with theory we were able to examine the socializing forces of our lives. In our initial view of the stories about our lives in the natural world, we focused on control, prediction, observation, and learning. Indeed some of the stories do reflect dominant cultural values relevant to science and nature, namely, that nature is a realm that can and should be controlled. Through memory-work, we excavated new meanings of stories that interrupt the dominant narrative of the natural world. We brought to the surface stories of what we learned through and with our families as well as stories that showed sensuous, creative, metaphoric, and other powerful connections to the natural world—in short, a different way of thinking about nature and science.

The discovery of our close and intimate connections with nature emerged from this work. In a cultural sense, we learned that we are socialized to forget these connections. They are not valued and, by and large, do not become a part of the stories we tell about our lives. Connections do not become a part of the social discourse. We come to accept and adhere to the prevailing idea that nature is separate from who we are. The question that undergirded this study was, "How have we been socialized to think about the natural world?" Now we can answer that we are taught not to value the connections or the ways in which we interrelate with nature that characterize our early experiences in the natural world. The result is a story of ourselves and the natural world that privileges the dominant discourse of separation advocated by traditional science. Through memory-work, we have interrupted that story and now better understand in both a theoretical and experiential way that we are part of nature. We described this experience of continuity in terms of the sensuous, the metaphoric, the creative, the family, and power.

Our Findings

In analysis we discovered that as children we engaged in what we have come to call personal science. This is a science that produces an embodied knowing with a strongly sensuous aspect so that we come to learn about air through the way it feels playing a clarinet, and we learn about earth through the way it feels and changes temperature on a hot summer day, or we know about water from the way our bodies plunge through it. When we began memory-work research, we were focused on the cognitive and the social in our memories. We were surprised to find this sensuous aspect and slow to recognize that our embodied experience led to a physical knowing of the natural world. We situated our knowledge in our sensory experience and recognized that it is an important component of both our personal and our professional science.

The elements and their metaphorical transformations helped us extend and further delineate our understanding of the sensuous connections and how these are shaped by our historical and cultural locations. Haug's (1987) injunction to look for cliché or metaphor brought us to notice the metaphors that are part of our memories, our discussions of those memories, and the language that we used in our analysis. In short as Lakoff and Johnson (1980) have observed, metaphor is essential to the fabric of our lives; it is part of the meaning we make of the world. Among the metaphors in our memories and their analysis, we found themes of fear, efficacy, relationship, and development and growth. Fears related to floods and tornadoes as well as fear of earth as in being buried alive or fear of not getting enough air shaped our connections to nature. Water and fire were two elements that challenged our feelings of efficacy. Competitiveness in building campfires, the "hardness" of water in tests of waterskiing and life-saving skills, and "*sweating*" over laboratory tasks were parts of memories in which we learned about success and failure. The classical elements provided metaphors for our relationships with others as in the sauna memory where Finnish roots and matriarchy came to the fore or in the fog memory in which fog obscured the future of a relationship. Finally, our memories of gardening, harvesting, and cultivating provided metaphors for our own growth and development.

In metaphoric associations with the elements, we discovered that sometimes we were drawn closer to nature and sometimes pushed away. In a gendered metaphor, girls are not supposed to get their hands "dirty," as shown in Sue's dirty leaves memory. Rae's memories illustrate the metaphorical link between water and life as well as the sustained and close connection that Rae feels with water. When we analyzed metaphors in our memories of science in school, we expected that we would gain insight into our socialization as apprentice scientists. Instead, and consistent with the developmental literature, we found core issues of competency and self-consciousness. For some of us, science in school was ultimately inviting (Rae's carrot water and Bunsen burner memories), and for others, it dampened our interest in natural science (Bell's sweaty palms memory). The desire to succeed and the fear of failure are powerful emotions in adolescence, and we believe that they strongly influence continuing interest in nature and traditional science.

The creativity in our memories or the ways in which we made new meaning for ourselves appeared to reinforce our recognition of our own science. Small spaces where we could engage in creative playing and pretending afforded close connections to the natural world. Rae's willow tree memory describes a place *"smelling like dirt and green."* Within these creative spaces we found multiple opportunities for imagining, problem solving, exploring, and experimenting. The fairy rock memory illustrates how connection to a natural space can be used to imagine new ways of being. Here, Rae uses water to make a place for swimming, moss to make a soft place, and sticks to make benches for the fairies. Cars and kitchens provided places where Bell let her arm fly in the wind, and Anna listened to her father's breath, both exploring and experimenting in air. We found flow (Csikszentmihalyi, 1996) as time was suspended in a challenging task, whether playing with the *"soft, yellow dirt"* or *"lost in allowing her limb to be played by the air."* Our memories also provided evidence of wonder when Sue watched the movement and listened to the sound of the candle carousel for hours or when Anna encountered magnesium. Our creative connections to nature as adolescents and adults were evident in Cele's watermelon memory with its adolescent challenges and the mystery of footprints in Sue's cat-prints memory.

Many varieties of power were evident in our work. We felt the power of earthquakes and tornadoes and felt powerful at the sight of San Francisco Bay and working in the quarry pond. We felt power when we engaged in our personal science and we were learning about the elements. We experienced power in nature when we were connected with other powerful women watching them build a box campfire or as we ventured into a formidable cave with a friend. Power *with* was more evident in our memories than power *over* although we found evidence of domination when Sue's father kicked a plate of sand back into the sandbox or when Anna's father yelled that her *"floozy"* skirt would *"burn like kerosene."* Our analysis revealed that under such circumstances, we appeared to forget our close connections to the natural world. As we grew older, we seemed to forget our personal science, even forget the power that we felt in nature. Through memory-work, we now tell a story of our work as scientists that includes power and the knowing gained through our personal science.

As family emerged as a theme in our memories, we saw some of the ways our values were learned and our relationships to the natural world were socially shaped. Our memories are evidence for ways in which we are shaped by Western culture's dominant discourse on nature as it emerges through our families. We remembered that our fathers controlled us, taught us, and allowed us to explore and play in the natural world. Some of us forgot our mothers and other powerful women who taught us to think about and experience nature. We appropriated the cultural view, but we also resisted it. We engaged in "clever disobedience" so that we could experience the natural world for ourselves. Our social worlds appeared to take precedence over our other close connections within the natural world as we grew into adolescence. In this division between the social and the natural, we seemed to mirror the cultural myth that we are able to divorce ourselves from the natural. Although many of us appeared to carry this belief into adulthood, memory-work shows us that the social and natural are inextricably intertwined.

Emerging Questions

Can the findings of this work be generalized? We would first reframe that question and ask, are our findings useful? We think personal science is quite useful, particularly in thinking about the context of schooling. The evidence that we found for our sensuous connections with the elements has direct implications for providing opportunities for related experiences in classrooms. For example, in science classrooms we engage students in activities that involve observing and predicting what will happen to water under various conditions, but distance from water is an implicit value in these activities. Along with observation and prediction, why not engage students in describing how water feels on their bodies under various conditions? What does it feel like to build something under water, to immerse your arm in cold water, in warm water? What does the warm air of a spring day feel like on your face?

Activities such as these might help students become familiar with a narrative of connection, which may be useful in learning. For example, they might augment the demonstrated value of a cognitive imagery exercise intended to improve understanding of the concepts of condensation and evaporation (Ewing and Mills, 1994). Students in this particular study were asked to remember the sensuous aspects of wading or sitting at the edge of a pond to strengthen their connection to the context of the water cycle in which the concepts were presented. We think it is valuable to sustain and nourish connections in response to a culture that often teaches us to forget them.

A second question is, are the findings of this study specific to our group? In some ways they must be, but we think our findings having to do with the sensuous, metaphor, creativity, power and the family illuminate our more general conclusions about personal science and science in school. We encourage other memory-work collectives to form and to examine some of the same questions we have posed. Such collectives could be identity based and formed along lines such as gender, ethnicity, class, or sexuality. Insights could be gained concerning the impact of identity on the relationship to nature.

The Validity of This Story

Our memory-work has helped us weave together a story that resurrects our interrelationship with the natural world. We now are learning that this is one of many stories we can tell based on the memories that we construct out of the continuous moments of our lives. Together we have become intimately aware through our memory-work that we tend to tell few rather than many stories. The stories that we tell often reveal and are constrained by the predominant values of the culture. These values effectively function to select the moments that we take from the time stream to tell a coherent narrative of our lives. We are aware of these values in varying degrees, but many are very familiar; we take them for granted. They have become invisible and sometimes oppressive influences in and around our lives. The values shape what we see and know, and we participate in perpetuating their influence in our culture. In this context, memory-work interrupts, making the invisible visible, making the familiar strange.

We are prompted to ask why we should take our stories seriously—or ask why these stories are more acceptable than previous stories we told. In short—can we justify these stories? Freeman (1993) suggests that we can judge or justify a narrative based on plausibility, and part of plausibility should be coherence—not necessarily in the sense of a tightly woven narrative, the kind that would set off alarms indicating cultural coherence, but rather that there is not blatant contradiction within the story. Freeman also suggests that the narrative should be fitting or sensible. However, most importantly for our purposes, the criterion of plausibility should not constrain us so much that we simply reproduce the culture-rather we should "remain open to entirely new forms of interpretation" (Freeman, 1993, p. 165). Nevertheless, Freeman cautions that interpretation will never escape "the scope of our own idioms and habits of thought" (1993, p. 165).

Freeman's exploration of memory is a solitary one. No others challenge him to break through the boundaries of the culture or the "idioms and habits of thought." The collective nature of memory-work, however, directly addresses this problem. The linking of self as researcher and self as participant in the process of memory-work forges connections that allow us to challenge one another to move beyond the idioms, the boundaries of our encultured space. While we cannot claim that we have moved past these boundaries, we can confirm that we see them more clearly as a collective. The collective has opened new possibilities for interpretation more readily than we could have individually. In this conclusion, we bring together the possibilities and some of the transformations that have occurred for us.

The Present View of the Past

A powerful cultural belief holds that the past defines our present lives. As a result of memory-work, we have found that the present also defines how we see the past. While we have noted linkages between the past and the present, our collective has uncovered for us novel ways of looking at those linkages.

When we first embarked on this project, it took an excruciatingly long time to move beyond the meanings we had inscribed on the narratives of our pasts, to begin to discern what Kermode (1979) calls the secrecy of our lives. This secrecy is what is invisible and comprises the aspects of our worlds that we do not value, pay attention to, or notice. The first time we began to see the outlines of the invisible occurred when Sue recalled that she did science in her mother's realm. Our pasts seemed not as rigid, not as inscribed. The possibility of other stories and other meanings became real.

Memory-work did not suddenly become easier; we still experienced frustration in our work with the memories. The secrecy was still there, but Sue's narrative, recalled in discussion of the sandbox memory, opened a door for us. Slowly new narratives began to emerge. We became accustomed to the discourse of reseeing and valuing things or attending to things we had not noticed previously. When we reexamined Rae's red rubber boots memory, we saw a child experiencing the sheer pleasure of feeling water stream over her boots rather than a child scientist wondering about how her boots altered the course of sticks in water flowing through a gutter. It seemed at times as if we existed in two worlds, the first being the world of inscribed meanings, the narratives of our lives that we were unconscious to, and a second world of new meanings and possibilities.

Questioning causal and linear sequences or descriptions has some wide-ranging implications. When we decided as a collective to include adolescent and adult memories along with our earliest memories of the elements, we assumed that we would uncover developmental stories. Questioning causal sequences, though, means that the very nature of development comes under scrutiny. Are the ideas of unfolding stages or sequences of behavior merely an artifact of a culture steeped in a discourse of linear causality? There is surely change, and in fact we can discuss our own development as a result of increasing experience with memory-work, the psychological and intellectual development that occurs with an expanding consciousness of our cultural context.

However, is there a natural course of development that implies that we move predictably from a lower to higher state? The multiple stories of our lives put into doubt that description of development as hierarchical. In fact, Morss (1996) argues that the idea of the child as a lesser form of an adult, an idea that permeates developmental theory, should be treated as a cultural story rather than a truth. Once viewed in this way, other stories of children and development become possible, and it becomes more difficult, for example, to say that one child is "less developed" than another or that one nation is "less developed" than another.

Connection to Nature: Personal Science

The sensuous experience of a cool stream, the vivid descriptions of hot and cold sand, and the sheer joy of feeling water running over red rubber boots express a wholeness or connectedness between body and nature. The dominant culture does not value embodied knowing, and hence we believe that this ability is not ordinarily explored or expanded as we grow.

The wholeness of embodied knowing has roots in the Romantic movement of the late eighteenth and early nineteenth centuries. Writers such as Thoreau, Emerson, and Muir articulated an organicism that Merchant (1980) describes as "a vital animating principle binding together the whole created world" (p. 100). In the present, we see our early experiences in the natural world as consistent with theories of ecofeminism (Lahar, 1993) and phenomenology (Bigwood, 1993).

We can speculate about the cultural implications for embracing a wholeness or connection between body and nature. Surely, the recognition that we are connected means that we would be less inclined to inflict harm. We are not immune to self-destruction, but we engage in such acts far less frequently than we engage in acts of destruction directed to things/objects we feel disengaged from, such as nonhuman nature.

The recognition of a personal science has clear implications for how we educate children, and it has implications for us as college educators. Indeed, Koch (1999) encourages us to locate our science selves, contending that all of us have engaged in some form of observation and prediction in nature, though this notion of a scientific self implies a distance between an observer and an object. Personal science in the way we have come to define it goes beyond observation and prediction and represents an interrelatedness between subject and object; the knowing that results is "written on the body" (Winterson, 1993).

It is a knowing that reunites intellect with feeling. It is a knowing that metaphorically connects a foggy night to a road not taken, as in Cele's adult fog memory, or a knowing that results from the simple pleasure of connection to air and water (as in Bell's arm in air memory, Cele's stream dam memory, and Anna's sauna memory). Perhaps a part of personal science is the joy and wonder we feel when we understand that we are connected to, part of, the natural world. At a molecular level, we are united and connected with the cycles and elements of the biosphere, even rocks

(Hayward, 1999). Giving voice to this connection, validating the wholeness, and animating the knowing that results are pleasing, joyful, and open up possibilities for an entirely different relationship to the natural world, which influences our professional science as well.

Transformations

Our stories have been transformed by memory-work. In Freeman's terms what has occurred is ideological critique and psychological development. New and critical ways of seeing the social and physical world changed us. The changes altered the way we now experience the world. In Haug's sense, it is yet another way in which theory is not only linked with experience but shapes it and is shaped by it.

What we have learned as a collective has begun to work its way into our practice as teachers and researchers. For those of us who are developmental researchers and engaged in longitudinal research, the stories we produce based on observation and interviews are constrained by not only the culture but also the disciplines within which we work. Our work then is open to an entirely new set of critical questions regarding the narratives participants provide and the sense that we make of those stories and of our observations. Several of us are engaged in longitudinal research and teaching with teachers and administrators in public school settings. We can now ask: What are the other stories these educators could tell about their practice and their growth? How do they become aware that there are other possible stories to tell? How does the researcher move beyond simply questioning the existing narrative to seriously considering other possible narratives? This is painstaking work, and memory-work suggests that we can open up the data beyond the boundaries of the discourse. But is this possible within a traditional paradigm in which subject and object are separated?

If one of our goals is to stop reproducing aspects of the culture that we believe are in need of change, then perhaps it is time to seek other possible stories that can be told about educational practice. In this case, the researcher would need to join with educators in conducting memory-work research. We have used portions of this method with teacher educators, graduate students in the sciences, students exploring definitions of science, and colleagues. We see multiple other adaptations of the method.

Certainly, the five of us will never look at science the same way again. Historically, science has imposed, through its definitions and methods, a belief system that may be contrary to our personal science. It has shaped, sometimes subtly and unconsciously, our interactions with nature. Science in school, we have seen, has the capacity to fuel or quash our interest in the natural world. Although we could tell other stories, we do believe there are connections between our personal science and professional lives, not as determinants out of the past but as important enriching links between personal experience and professional mastery and power in nature today. Now we can define ourselves as scientists through our personal science as well. For example, Bell has struggled with doubts related to the close connections she often experiences with the "subjects" of her research. She understands now that she and they are both subject and object and that distance between the two is illusory. As a result of this work, Rae recognizes and acknowledges the close connections with the objects she studies, having believed that science excludes close connection. We might suppose that most scientists may not have close connections with the things they study because such a relationship is rarely talked about in the literature or professional meetings, Keller's (1983) account of Barbara McClintock's work notwithstanding. If it is rarely or never talked about, it cannot become a part of the discourse. Rae now recognizes and acknowledges a language of connection that has been obscured and not valued. Furthermore, as a result of this work, Anna now uses some aspect of memory-work in every course she teaches.

The consciousness that results from this work must be practiced with vigilance. It is easy to fall into the old ways of seeing because we continue to live in a social world that constantly reinforces old boundaries and meanings. Nevertheless, the interruptions occasioned by this work have changed our narratives. We will continue to tell very different stories.

Illustrations

Figure 1.

Growing Rocks. Spring and time to prepare the garden soil, she wonders how the boulders can reappear year after year.

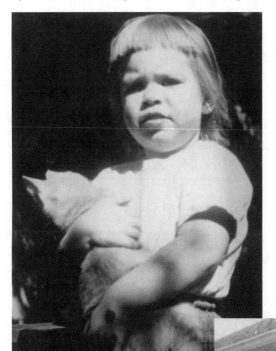

Figure 2.

Cat Prints (above). At ease with a cat decades before she cleaned up the cat prints.

Dirty Leaves (right). By the fence next to the basement steps, which must be cleared of the dreaded rotting leaves every fall.

Figure 3.

Family Planting. Planting helpers climb the big slide, another part of growing a family. (Photo by Helen Jordan)

Figure 4.

Fairy Rock. The fairy rock awaits landscaping under the forsythia bush. (Photo by John Steffens)

Figure 5.

San Francisco Bay. Savoring the freedom and possibility of this new place.

Figure 6.

Junior Prom. After riding their motorcycles to choose the trees, the girls cut them and bring them back in the station wagon.

Appendix A

Appendix A: List of Memories

	Sue	Anna	Bell	Rae	Cele
Childhood Memories					
Fire 3/3/94	candle carousel	spring burn	matches	candle flame; cigarette; paper balloon	chocolate milk
Water 3/3/94	puddle	sauna	club pool	red boots; fairy rock	stream dam
Air 4/21/94	up in air	breathing	arm in air	swing; thunderstorm; bronchitis	underwater swim; cave
Earth 5/5/94	sandbox	muck	moving the graves	digging; alley dirt	dirt potatoes
Tree 5/25/94	seeds	willow flute	tree house	willow tree; dead snake; carriage; empty lots	red sugar maple; calf
Adolescent Memories					
Fire 8/30/94	campfire	magnesium	matchless fire; sweaty palms	Bunsen burner; fire alone	box campfire
Water 9/13/94	bikini	waterskiing	WSI	carrot water; canoeing	watermelon
Air 9/27/94	smoking; Ferris wheel	bogeyman	tower picture	soft coal; poem; asthma; clarinet	tornado
Earth 11/2/94	ivy	motorcycle garden	growing rocks	farm kiss	chat piles
Tree 11/2/94	dirty leaves	Jr. Prom	glass factory	arboretum	car tree
Adult Memories					
Fire 11/2/94	fireplace flue	naboom wood	candlelight vigil	Halloween baby	family fire
Water 11/30/94	flooding	headache	San Francisco Bay	pond wading; quarry pond	glacier
Air 11/30/94	balloons	leaving; snow	dunes ranger	flying	fog
Earth 12/14/94	cat prints	poor potatoes	earthquake	family planting	grape miracle
Tree 12/14/94	crooked Xmas tree	almonds	up in a tree	beds and trees	allergies

Appendix B

Appendix B: Memory Locations, Trigger, Age, and Author

Chapter	Memory	Trigger	Age	Author	Full Text	Page
1. Introduction	sweaty palms	Fire	Adolescent	Bell	X	1
	sandbox	Earth	Child	Sue		3
3. Memory and Memory-work	paper balloon	Fire	Child	Rae	X	33
	growing rocks	Earth	Adolescent	Bell	X	34
4. Making Sense	almonds	Tree	Adult	Anna	X	47
	glacier	Water	Adult	Cele	X	49
	sauna	Water	Child	Anna	X	50
	Ferris wheel	Air	Adolescent	Sue	X	50
	clarinet	Air	Adolescent	Rae	X	51
	bikini	Water	Adolescent	Sue	X	52
	box campfire	Fire	Adolescent	Cele	X	52
	breathing	Air	Child	Anna	X	54
	tornado	Air	Adolescent	Cele	X	55
	dirty leaves	Tree	Adolescent	Sue	X	57
	earthquake	Earth	Adult	Bell	X	57
	muck	Earth	Child	Anna	X	58
	quarry pond	Water	Adult	Rae	X	59
	bogeyman	Air	Adolescent	Anna	X	60
	dirt potatoes	Earth	Child	Cele	X	60
	sandbox	Earth	Child	Sue	X	61
	digging	Earth	Child	Rae	X	62
	arm in air	Air	Child	Bell	X	62
5. Metaphor	sweaty palms	Fire	Adolescent	Bell		68
	tree house	Tree	Child	Bell		69
	underwater swim	Water	Child	Cele		71
	stream dam	Water	Child	Cele		71
	chocolate milk	Fire	Child	Cele		71
	chat piles	Earth	Adolescent	Cele	X	72
	matches	Fire	Child	Bell		73
	matchless fire	Fire	Adolescent	Bell		73
	WSI	Water	Adolescent	Bell		74
	carrot water	Water	Adolescent	Rae		74
	sauna	Water	Child	Anna		75
	fog	Air	Adult	Cele		75
	glass factory	Tree	Adolescent	Bell	X	76
	car tree	Tree	Adolescent	Cele	X	77
	growing rocks	Earth	Adolescent	Bell		78
	dirty leaves	Tree	Adolescent	Sue		78
	family planting	Earth	Adult	Rae	X	78
	smoking	Air	Adolescent	Sue		78
	up in a tree	Tree	Adult	Bell		79
	candlelight vigil	Fire	Adult	Bell		79
	sweaty palms	Fire	Adolescent	Bell		80
	Bunsen burner	Fire	Adolescent	Rae	X	81
	magnesium	Fire	Adolescent	Anna		83

(continued)

Chapter	Memory	Trigger	Age	Author	Full Text	Page
6. Making New Meaning	watermelon	Water	Adolescent	Cele	X	86
	cat prints	Earth	Adult	Sue	X	87
	willow tree	Tree	Child	Rae	X	89
	digging	Earth	Child	Rae		89
	fairy rock	Water	Child	Rae	X	90
	fairy rock rewrite	Water	Child	Rae	X	90
	sandbox	Earth	Child	Sue		92
	allergies	Tree	Adult	Cele		93
	flying	Air	Adult	Rae		93
	alley dirt	Earth	Child	Rae		94
	arm in air	Air	Child	Bell		94
	clarinet	Air	Adolescent	Rae		95
	candle carousel	Fire	Child	Sue	X	95
	magnesium	Fire	Adolescent	Anna		96
	glacier	Water	Adult	Cele		96
	grape miracle	Earth	Adult	Cele	X	96
	earthquake	Earth	Adult	Bell		97
	balloons	Air	Adult	Sue		97
	San Francisco Bay	Water	Adult	Bell		97
7. Family Landscapes	sandbox	Earth	Child	Sue		100
	stream dam rewrite	Water	Child	Cele	X	100
	spring burn rewrite	Fire	Child	Anna	X	102
	willow flute	Tree	Child	Anna	X	103
	arm in air rewrite	Air	Child	Bell	X	103
	breathing	Air	Child	Anna		104
	box campfire	Fire	Adolescent	Cele		107
	club pool	Water	Child	Bell	X	108
	matches	Fire	Child	Bell		109
	dirt potatoes rewrite	Earth	Child	Cele	X	110
	cave	Air	Child	Cele	X	110
	motorcycle garden	Earth	Adolescent	Anna	X	111
	sauna	Water	Child	Anna		112
	San Franscisco Bay	Water	Adult	Bell		114
	candlelight vigil	Fire	Adult	Bell		114
8. Power of Girls	earthquake	Earth	Adult	Bell		118
	car tree	Tree	Adolescent	Cele		118
	up in air	Air	Child	Sue		118
	up in air rewrite	Air	Child	Sue	X	119
	sandbox	Air	Child	Sue		119
	junior prom	Tree	Adolescent	Anna		121
	dunes ranger	Air	Adult	Bell		121
	carrot water	Water	Adolescent	Rae		121
	box campfire	Fire	Adolescent	Cele		122
	club pool	Water	Child	Bell		122
	flying	Air	Adult	Rae		123
	beds and trees	Tree	Adult	Rae		123
	San Francisco Bay	Water	Adult	Bell		123
	candle flame	Fire	Child	Rae	X	124

(continued)

Chapter	Memory	Trigger	Age	Author	Full Text	Page
8. Power of Girls	cigarette	Fire	Child	Rae	X	124
	paper balloon	Fire	Child	Rae		124
	thunderstorm	Air	Child	Rae		124
	arm in air	Air	Child	Bell		126
	matchless fire	Fire	Adolescent	Bell		126
	WSI	Water	Adolescent	Bell		127
	waterskiing	Water	Adolescent	Anna		128
	Bunsen burner	Fire	Adolescent	Rae		128
	magnesium	Fire	Adolescent	Anna		128
	family planting	Earth	Adult	Rae		128

References

Abram, David. (1996). *The spell of the sensuous: Perception and language in a more-than-human world*. New York: Vintage Books.

Allured, Janet. (1992, Spring). Women's healing art: Domestic medicine in the turn-of-the-century Ozarks. *Gateway Heritage, 12,* 20–31.

Amabile, Teresa. (1983). *The social psychology of creativity*. New York: Springer.

American Association of University Women. (1992). *How schools shortchange girls: A study of major findings on girls and education*. Washington, DC: AAUW Education Foundation and National Education Association.

Anderson, John R. (1983). *The architecture of cognition*. Cambridge: Harvard University Press.

Anderson, Margaret L. (1990). Changing the curriculum in higher education. In Clifton F. Conrad & Jennifer Grant Haworth (eds.), *Curriculum in transition: Perspectives on the undergraduate experience* (pp. 119–144), ASHE Reader Series. Needham Heights, MA: Ginn Press.

Arendt, Hannah. (1972). *Crises of the republic: Lying in politics, civil disobedience on violence, thoughts on politics, and revolution*. New York: Teachers College Press.

Atkinson, Richard C., & Shiffrin, Rose M. (1968). Human memory: A proposed system and its control processes. In Kenneth W. Spense (ed.), *Psychology of learning and motivation: Advances in research and theory* (vol. 2, pp. 89–195). New York: Academic Press.

Audesirk, Terry, & Audesirk, Gerald. (1996). *Biology: Life on earth* (4th ed.). Upper Saddle River, NJ: Prentice Hall.

Ayman-Nolley, Saba. (1992). Vygotsky's perspective on the development of imagination and creativity. *Creativity Research Journal, 5*(1), 77–86.

Bachelard, Gaston. (1964). *The psychoanalysis of fire* (Alan C. M. Ross, trans.). Boston: Beacon Press. (Original work published 1938)

Bachelard, Gaston. (1988). *Air and dreams: An essay on the imagination of movement*. Dallas: The Dallas Institute Publications. (Original work published in 1943)

Bakan, David. (1966). *The duality of human existence: An essay on psychology and religion*. Chicago: Rand McNally & Co.

Baltes, Paul B., Reese, Hayne W., & Nesselroade, John R. (1977). *Lifespan developmental psychology: Introduction to research methods*. Belmont, CA: Wadsworth Publishing.

Barber, Leslie A. (1995). U.S. women in science and engineering, 1960–1990: Progress toward equity? *Journal of Higher Education, 66*(2), 213–234.

Barinaga, Marcia. (1993). Feminists find gender everywhere in science. *Science, 260*(5106), 392–393.

Barr, Jean, & Birke, Lynda. (1998). *Common science? Women, science, and knowledge.* Bloomington, IN: Indiana University Press.

Bateson, Mary Catherine. (1989). *Composing a life.* New York: Penguin Books.

Bateson, Mary Catherine. (1991). *Our own metaphor.* Washington, DC: Smithsonian Institution Press.

Bateson, Mary Catherine. (2000). *Full circles, overlapping lives: Culture and generation in transition.* New York: Random House.

Belenky, Mary Field, Clinchy, Blythe McVicker, Goldberger, Nancy Rule, & Tarule, Jill Mattuck. (1986). *Women's ways of knowing: The development of self, voice, and mind.* New York: Basic Books.

Bigwood, Carol. (1993). *Earth muse: Feminism, nature, and art.* Philadelphia: Temple University Press.

Birke, Lynda. (1999). *Feminism and the biological body.* New Brunswick, NJ: Rutgers University Press.

Bjork, Elizabeth Ligon, & Bjork, Robert A. (1996). *Memory.* San Diego, CA: Academic Press.

Brown, Lyn Mikel. (1998). *Raising their voices: The politics of girls' anger.* Cambridge, MA: Harvard University Press.

Brown, Lyn Mikel, & Gilligan, Carol. (1992). *Meeting at the crossroads.* Cambridge, MA: Harvard University Press.

Brunn, Michael. (1994, April). *Ethnohistories: Learning through the stories of life experiences.* Paper presented at the annual meeting of the American Educational Research Association, New Orleans (American Indian, Alaska Native Education SIG).

Brunner, C. Cryss, & Schumaker, Paul. (1998). Power and gender in the "new view" public schools. *Policy Studies Journal, 26*(1), 30–45.

Bryman, Alan. (1988). *Quantity and quality in social research.* London: Unwin Hyman.

Byrne, Eileen M. (1993). *Women and science: The snark syndrome.* London: The Falmer Press.

Cameron, Julia. (1992). *The artist's way: A spiritual path to higher creativity.* New York: Tarcher/Putnam Book.

Campbell, Neil A. (1993). *Biology* (3rd ed.). New York: Benjamin Cummings.

Campbell, Rebecca, & Schram, Pamela J. (1995). Feminist research methods: A content analysis of psychology and social science textbooks. *Psychology of Women Quarterly, 19*(1), 85–106.

Carter, Rita. (1998). *Mapping the mind.* Berkeley: University of California Press.

Cataldi, Sue L. (1993). *Emotion, depth, and flesh: A study of sensitive space.* Albany, NY: State University of New York Press.

Chambers, D. W. (1983). Stereotypic images of the scientist: The draw-a-scientist test. *Science Education, 67,* 255–265.

Civian, Janet T., Rayman, Paula M., Brett, Belle, & Baldwin, Lawrence M. (1997). *Path-*

ways for women in the sciences. The Wellesley report part II. Wellesley, MA: Wellesley College, Center for Research on Women.

Clandinin, D. Jean, & Connelly, F. Michael. (1995, April). *Storying and restorying ourselves: Narrative and reflection.* Paper presented at the annual meeting of the American Educational Research Association, San Francisco.

Cobb, Edith. (1959). The ecology of imagination in childhood. *Daedalus, 88,* 537–548.

Cole, A. L. (1991). Interviewing for life history: A process of ongoing negotiation. In Ivor F. Goodson & John Marshall Mangan (eds.), *Qualitative educational research studies: Methodologies in transition, Occasional Papers, 1* (pp.185–208). London, ONT: Research Unit on Classroom Learning and Computer Use in Schools.

Colwell, Rita. (1999). Preface. In Elga Wassermann, *The door in the dream: Conversations with eminent women in science* (pp. ix–xii). Washington, DC: Joseph Henry Press.

Connelly, F. Michael, & Clandinin, D. Jean. (1990). Stories of experience and narrative inquiry. *Educational Researcher, 19*(5), 2–14.

Crawford, June, Kippax, Susan, Onyx, Jenny, Gault, Una, & Benton, Pam. (1992). *Emotion and gender: Constructing meaning from memory.* Newbury Park, CA: Sage.

Csikszentmihalyi, Mihaly. (1996). *Creativity: Flow and the psychology of discovery and invention.* New York: HarperCollins.

Csikszentmihalyi, Mihaly. (1997). *Finding flow: The psychology of engagement of everyday life.* New York: Basic Books.

Degler, Carl. (1980). *At odds: Women and the family in America from the revolution to the present.* New York: Oxford.

Didion, Catherine J. (1993). Making teaching environments hospitable for women in science. *Journal of College Science Teaching, 23,* 82–83.

Dodgson, Charles Lutwidge. (1966). *The hunting of the snark; an agony in eight fits, by Lewis Carroll.* New York: Pantheon Books.

Egan, Kieran. (1997). *The educated mind: How cognitive tools shape our understanding.* Chicago: University of Chicago.

Eichler, Margrit. (1988). *Nonsexist research methods.* Boston: Allyn and Unwin.

Eisenhart, Margaret A., & Finkel, Elizabeth. (1998). *Women's science: Learning and succeeding from the margins.* Chicago: University of Chicago Press.

Engel, Susan. (1999). *Context is everything: The nature of memory.* New York: W. H. Freeman and Co.

Evans, Mary. (1999). *Missing persons: The impossibility of auto/biography.* London: Routledge.

Evetts, Julia. (1996). *Gender and career in science and engineering.* London: Taylor & Francis.

Ewing, Margaret S., & Mills, Terence J. (1994). Water literacy in college freshmen: Could a cognitive imagery strategy improve understanding? *Journal of Environmental Education, 25,* 36–40.

Fausto-Sterling, Anne. (1997). Women's studies and science. *Women's Studies Quarterly, 25,* 183–189. (Original work published 1980)

Follett, Mary Parker. (1924). *Creative experience.* New York: Longmans, Green and Co.

Freeman, Mark. (1993). *Rewriting the self: History, memory, narrative.* London: Routledge.

Freire, Paulo. (1985). *The politics of education.* South Hadley, MA: Bergin and Garvey.

Garfield, Eugene. (1986). The metaphor-science connection. *Current Contents, 42,* 3–10.

Gibson, James J. (1986). *The ecological approach to visual perception*. Hillsdale, NJ: Lawrence Erlbaum.

Gilligan, Carol. (1982). *In a different voice: Psychological theory and women's development*. Cambridge, MA: Harvard University Press.

Goldstein, Lisa S. (1999). The relational zone: The role of caring relationships in the co-construction of mind. *American Educational Research Journal, 36*, 647–673.

Graziano, Anthony M., & Raulin, Michael L. (1997). *Research methods: A process of inquiry*. New York: Longman.

Green, Thomas F. (1979). Learning without metaphor. In Andrew Ortony (ed.), *Metaphor and thought* (pp. 462–473). Cambridge: Cambridge University Press.

Grumet, Madeleine R. (1990). Retrospective: Autobiography and the analysis of educational experience. *Cambridge Journal of Education, 20*(3), 321–325.

Harding, Sandra. (1992). How the women's movement benefits science: Two views. In Gill Kirkup & Laurie Smith Keller (eds.), *Inventing women: Science, technology and gender* (pp. 57–72). Cambridge, MA: Blackwell Publishers (Polity Press).

Harter, Susan. (1999). *The construction of the self: A developmental perspective*. New York: The Guilford Press.

Harter, Susan, Bresnick, Shelly, Bouchey, Heather A., & Whitesell, Nancy R. (1997). The development of multiple role-related selves during adolescence. *Development and Psychopathology, 9*, 835–853.

Haug, Frigga. (ed.). (1987). *Female sexualization: A collective work of memory* (Erica Carter, trans.). Towbridge, Wiltshire, UK: Dotesios Ltd.

Hayward, Jeremy. (1999). Unlearning to see the sacred. In Steven Glazer (ed.), *The heart of learning: Spirituality in education* (pp. 61–76). New York: Jeremy P. Tarcher/Putnam.

Helgesen, Sally. (1990). *The female advantage: Women's ways of leadership*. New York: Doubleday.

Hermanowicz, Joseph C. (1998). *The stars are not enough: Scientists—Their passions and professions*. Chicago: University of Chicago Press.

Hill, Susan T. (1992). *Undergraduate origins of recent science and engineering doctorate recipients* (Special Report NSF 92–332). Washington, DC: The National Science Foundation.

Hoagland, Sarah Lucia. (1988). *Lesbian ethics: Toward new value*. Palo Alto: Institute of Lesbian Studies.

Holden, Constance. (1998). Defining "science" for the people. *Science, 280*(5364), 663.

Hubbard, Ruth. (1988). Science, facts, and feminism. *Hypatia, 3*(1), 5–17.

Jagger, Alison M. (1989). Love and knowledge: Emotion in feminist epistemology. In Alison M. Jagger & Susan R. Bordo (eds.), *Gender/body/knowledge: Feminist reconstructions of being and knowing* (pp. 145–171). New Brunswick, NJ: Rutgers University Press.

Jarvis, Tina. (1996). Examining and extending young children's views of science and scientists. In Lesley H. Parker, Leonie J. Rennie, & Barry J. Fraser (eds.), *Gender, science and mathematics: Shortening the shadow* (pp. 29–40). Boston: Kluwer Academic Publishers.

Jayaratne, Toby Epstein, & Stewart, Abigail J. (1991). Quantitative and qualitative methods in the social sciences: Current feminist issues and practical strategies. In Mary Margaret Fonow & Judith A. Cook (eds.), *Beyond methodology: Feminist scholarship as lived research* (pp. 85–106). Bloomington: Indiana University Press.

Johnson, Mark. (1987). *The body in the mind.* Chicago: University of Chicago Press.

Kanter, Rosabeth Moss. (1977). *Men and women of the corporation.* New York: Basic Books.

Keller, Evelyn Fox. (1983). *A feeling for the organism: The life and work of Barbara McClintock.* San Francisco: Freeman and Company.

Keller, Evelyn Fox. (1985). *Reflections on gender and science.* New Haven, CT: Yale University Press.

Kermode, Frank. (1979). *The genesis of secrecy.* Cambridge, MA: Harvard University Press.

Kippax, Susan, Crawford, June, Waldby, C., & Benton, Pam. (1990). Women negotiating heterosex: Implications for AIDS prevention. *Women's Studies International Forum,* 13(2), 533–43.

Koch, Janice. (1999). *Science stories: Teachers and children as science learners.* Boston: Houghton Mifflin.

Kotre, John. (1984). *Outliving the self.* Baltimore: Johns Hopkins University Press.

Kreisberg, Seth. (1992). *Transforming power: Domination, power, empowerment, and education.* Albany, NY: State University of New York Press.

Kuhn, Thomas S. (1962). *The structure of scientific revolutions.* Chicago: The University of Chicago Press.

Kuhn, Thomas S. (1979). Metaphor and science. In Andrew Ortony (ed.), *Metaphor and thought* (pp. 409–419). Cambridge, MA: Cambridge University Press.

Lahar, Stephanie. (1993). Ecofeminism and the politics of reality. In Geta Claire Gaard (ed.), *Ecofeminism: Women, animals and nature* (pp. 91–117). Philadelphia: Temple University Press.

Lakoff, George, & Johnson, Mark. (1980). *Metaphors we live by.* Chicago: University of Chicago Press.

Lakoff, George, & Turner, Mark. (1989). *More than cool reason: A field guide to poetic metaphor.* Chicago: University of Chicago Press.

Legerstee, M. (1991). Changes in the quality of infant sounds as a function of social and nonsocial stimulation. *First Language, 11,* 327–343.

Leslie, Larry L., McClure, Gregory T., & Oaxaca, Ronald L. (1998). Women and minorities in science and engineering. *Journal of Higher Education,* 69(3), 239–276.

Lewis, Richard. (1998). *Living by wonder: The imaginative life of childhood.* New York: Parabola Books in Association with Touchstone Center Publications.

Lutz, D. (1994). What is a scientist? *American Scientist, 82,* 575.

Lyotard, Jean Francois. (1984). *The postmodern condition: A report on knowledge* (G. Bennington & B. Massum, trans.). Minneapolis: University of Minnesota Press.

Macilwain, Colin. (1998). Physicists seek definition of "science." *Nature, 392* (6679), 849.

Maynard, Mary. (1997). Revolutionizing the subject: Women's studies and the sciences. In Mary Maynard (ed.), *Science and the construction of women* (pp. 1–14). London: UCL Press.

Mayr, Ernst. (1997). *This is biology: The science of the living world.* Cambridge, MA: Harvard University Press.

Merchant, Carolyn. (1980). *The death of nature: Women, ecology and the scientific revolution.* San Francisco: Harper and Row.

Miller, Jean B. (1982). *Women and power.* Work in progress #82-01. Wellesley, MA: Stone Center Working Papers Series.

Miller, Jean B. (1986). *Toward a new psychology of women* (2nd ed.). Boston: Beacon Press.

Miller, Jean Baker, & Stiver, Irene Pierce. (1997). *The healing connection: How women form relationships in therapy and in life.* Boston: Beacon.

Morss, John R. (1996). *Growing critical: Alternatives to developmental psychology.* New York: Routledge.

Nabhan, Gary Paul, & Trimble, Stephen. (1994). *The geography of childhood: Why children need wild places.* Boston: Beacon.

Nachmanovitch, Stephen. (1990). *Free play: The power of improvisation in life and the arts.* New York: Putnam.

Neumann, William Lawrence. (1997). *Social research methods: Qualitative and quantitative approaches.* Boston: Allyn and Bacon.

Norman, Donald. (1970). Comments on the information structure of memory. *Acta Psychologica, 33,* 292–303.

Overton, Willis F. (1998). Developmental psychology: Philosophy, concepts and methodology. In William Damon and Richard M. Lerner (eds.), *Handbook of child psychology,* 5th ed. (pp. 107–188). New York: John Wiley.

Oxford English Dictionary, 2nd ed. (1989). Oxford: Oxford University Press.

Petrie, Hugh G. (1979). Metaphor and learning. In Andrew Ortony (ed.), *Metaphor and thought* (pp. 438–461). Cambridge, MA: Cambridge University Press.

Piirto, Jane. (1998). *Understanding those who create.* Scottsdale, AZ: Gifted Psychology Press.

Porter, Theodore. (1995). *Trust in numbers: The pursuit of objectivity in science and public life.* Princeton, NJ: Princeton University Press.

Rich, Adrienne. (1979). Toward a woman-centered university. In *On lives, secrets, and silence: Selected prose* (pp. 125–155). New York: W. W. Norton & Co.

Rich, Adrienne. (1984). *The fact of a doorframe: Poems selected and new (1950–1984).* New York: W. W. Norton & Company.

Rogoff, Barbara. (1990). *Apprenticeship in thinking: Cognitive development in social context.* New York: Oxford University Press.

Rogoff, Barbara. (1995). Observing sociocultural activity on three planes: Participatory appropriation, guided participation, and apprenticeship. In James V. Wertsch, P. Del Rio, & A. Alvarez (eds.), *Sociocultural studies of mind* (pp. 139–164). Cambridge, MA: Cambridge University Press.

Rogoff, Barbara, Mistry, Jayanthi, Concu, Artin, & Mosier, Christine. (1993). Guided participation in cultural activity by toddlers and caregivers. *Monograph of the Society for Research in Child Development, 58*(8).

Rose, Steven. (1998). *Lifelines: Biology beyond determinism.* New York: Oxford University Press.

Rosener, Judy B. (1990). Ways women lead. *Harvard Business Review, 69*(6), 110–125.

Sacks, Oliver. (1995). *An anthropologist on Mars.* New York: Knopf.

Sadker, Myra, & Sadker, David. (1994). *Failing at fairness: How our schools cheat girls.* New York: Simon & Schuster.

Schmuck, Patricia A. (1999). Foreword. In C. Cryss Brunner (ed.), *Sacred dreams: Women and the superintendency* (pp. ix–xiii). Albany, NY: State University of New York Press.

Schratz, Michael, & Walker, Rob, with Schratz-Hadwich, Barbara. (1995). Collective memory-work: The self as a re/source for re/search. In Michael Schratz & Rob Walker, *Research as social change: New opportunities for qualitative research* (pp. 39–64). London: Routledge.

Scott, Anne. (1997). The knowledge in our bones: Standpoint theory, alternative health and the quantum model of the body. In Mary Maynard (ed.), *Science and the construction of women* (pp. 106–125). London: UCL Press.

Seymour, Elaine, & Hewitt, Nancy M. (1997). *Talking about leaving: Why undergraduates leave the sciences*. Boulder, CO: Westview Press.

Smolucha, Francine. (1992). A reconstruction of Vygotsky's theory of creativity. *Creativity Research Journal, 5*(1), 49–67.

Sonnert, Gerhard, & Holton, Gerald. (1995a). *Gender differences in science careers: The project access study*. New Brunswick, NJ: Rutgers University Press.

Sonnert, Gerhard, & Holton, Gerald. (1995b). *Who succeeds in science? The gender dimension*. New Brunswick, NJ: Rutgers University Press.

Spanier, Bonnie. (1995). *Im/partial science: Gender ideology in molecular biology*. Bloomington, IN: Indiana University Press.

Starhawk. (1987). *Truth or dare*. San Francisco: Harper and Row.

Strahler, Arthur N. (1992). *Understanding science: An introduction to concepts and issues*. Buffalo, NY: Prometheus Books.

Sutton-Smith, Brian. (1997). *The ambiguity of play*. Cambridge, MA: Harvard University Press.

Taylor, Jill McLean, Gilligan, Carol, & Sullivan, Amy M. (1995). *Between voice and silence: Women and girls, race and relationships*. Cambridge, MA: Harvard University Press.

Tobias, Sheila. (1992). Women in Science—Women and science. *Journal of College Science Teaching, 21*(5), 26–28.

Trevarthen, Colwyn. (1979). Communication and cooperation in early infancy: A description of primary intersubjectivity. In M. Bullowa (ed.), *Before speech: The beginning of interpersonal communication* (pp. 321–347). Cambridge, MA: Cambridge University Press.

Tulving, Endel. (1972). Episodic and semantic memory. In Endel Tulving (ed.), *Organization of memory* (pp. 382–403). New York: Academic Press.

Tulving, Endel. (1983). *Elements of episodic memory*. New York: Oxford University Press.

von Hornbostel, Erich M. (1927). The unity of the senses. *Psyche, 7*(28), 83–89.

von Oech, Roger. (1983). *A whack on the side of the head: How to unlock your mind for innovation*. New York: Warner.

Vygotsky, Lev Semenovich. (1978). *Mind in society: The development of higher psychological processes*. Cambridge, MA: Harvard University Press.

Vygotsky, Lev Semenovich. (1981). The genesis of higher mental functions. In James V. Wertsch (ed.), *The concept of activity in Soviet psychology* (pp. 144–188). Armonk, NY: Sharpe.

Vygotsky, Lev Semenovich. (1986). *Thought and language*. Cambridge, MA: Massachusetts Institute of Technology Press. (Original work published 1934)

Vygotsky, Lev Semenovich. (1990). Imagination and creativity in childhood. *Soviet Psychology, 28*(1), 84–96. (Original work published 1930)

Vygotsky, Lev Semenovich. (1994). Imagination and creativity of the adolescent. In René van der Veer & Jaan Valsiner (eds.), *The Vygotsky Reader* (pp. 266–288). Cambridge, MA: Blackwell.

Walker, Alice. (1992). *Possessing the secret of joy*. New York: Harcourt Brace Jovanovich.

Welty, Eudora. (1984). *One writer's beginnings*. Cambridge, MA: Harvard University Press.

Wertsch, James V. (1991). *Voices of the mind: A socio-cultural approach to mediated action*. Cambridge, MA: Harvard University Press.

Winnicott, Donald Woods. (1971). *Playing and reality*. London: Tavistock.

Winterson, Jeanette. (1993). *Written on the body*. New York: Vintage Books.

Witherell, Carol, & Noddings, Nel. (eds.). (1991). *Stories lives tell: Narrative and dialogue in education*. New York: Teachers College Press.

Wynne, B. E. (1991). Knowledges in context. *Science, Technology and Human Values*, 16, 111–21.

Index

A Feeling for the Organism (Keller), 17
Abram, D., 45, 49, 53, 56
affordance, 64, 82, 86, 91
agency, 117
Allured, J., 47
Amabile, T., 89
amalgams, 35
American Association of University Women, 22, 31
American Physical Society, 15
Anderson, J. R., 25
apprenticeship, 117, 131
    in families, 123–25
    multiple planes of, 125
    and power, 130
Arendt, H., 120
Atkinson, R., 39
Audesirk, G., 66
Audesirk, T., 66
autokoenony, 127
autotelic activity, 94
Ayman-Nolley, S., 92

Bachelard, G., 73, 79, 81, 90
    "clever disobedience," 107
Bakan, D., 117–18
Baldwin, L. M., 21
Baltes, P. B., 15
Barber, L. A., 20
Barinaga, M., 17
Barnes, E., 24
Barr, J., 19, 20, 47
Bateson, M. C., 23, 82, 88
Belenky, M. F., 19

Benton, P., 2
Bigwood, C., 61, 65, 82, 138
biodiversity, 23
Birke, L., 19, 20, 47, 63, 64
Bjork, E. L., 39
Bjork, R. A., 39
Brett, B., 21
Brown, L. M., 31, 119, 120
Brunn, M., 29
Brunner, C. C., 116, 120
Bryman, A., 18
Byrne, E. M., 16, 20

Cameron, J., 89
Campbell, N., 24
Campbell, R., 17
Carroll, L., 20
Carter, R., 40
Cataldi, S. L., 5, 64, 83, 86
Chambers, D. W., 13, 14
children
    experiencing nature, 24, 89
    perceptions of science, 13, 14–15
    and small spaces, 89, 90, 91, 134
Civian, J. T., 21, 22
Clandinin, D. J., 28, 29
"clever disobedience," 73, 107–12, 135
Clinchy, B. M., 19
coactive power, 120
Cobb, E., 3, 64
cognitive recall, 33
Cole, A. L., 28
Colwell, R., 22
Composing a Life (Bateson), 23

Concu, A., 5
Connelly, F. M., 28, 29
constructing knowledge, 63
construction of self, 127
constructivism, 4
context stripping, 47
continuity with nature, 132
control. *See* power
Crawford, J., 2, 3, 25, 26, 27, 28, 31, 34, 37,
    71, 109, 110, 111, 119
creativity, 7, 134
    creative spaces, 86–87, 91
    exploration and experimentation as a form
        of, 91–92
    and flow, 94–95
    and imagination, 90–91
    and new meaning, 85
    and personal science, 92
    problem solving as a form of, 92
    and wonder, 95–98
Csikszentmihalyi, M., 93, 94, 134
    autotelic activity, 94
culture and nature, 98

Das Argument, 4
Degler, C., 88
developmental stories, 137–38
Didion, C. J., 20, 21
distance from the elements, myth of, 80–82
distance from nature, 64, 132
Dodgson, C. L., 20

Egan, K., 98
Eichler, M., 18
Eisenhart, M. A., 22, 23, 24
elements of nature. *See* nature
Emerson, R. W., 138
emotional depth and physical distance, 5
Engel, S., 40
Evans, M., 29
Evetts, J., 23
Ewing, M. S., 135

family, 6, 8, 134–35
    chosen, 113–14
    and connections to nature, 113
    and control, 100
    and differences, 112–13

fathers' influence, 99–105
    mothers' influence, 105–7
    and rule-knowing, 100, 110
fantasy, 90
Fausto-Sterling, A., 20
Finkel, E., 22, 23, 24
flow, 94–95, 134
Follett, M. P., 116, 120
Frauenformen, 4
Freeman, M., 40, 41, 136, 139
Freire, P., 47, 65

Garfield, E., 82
Gault, U., 2
gender training, 119
Gibson, J. J., 5, 64
Gilligan, C., 31, 119
girls
    anxiety about laboratory science, 82
    compromising power, 120
    power of, 116
    and science, 20
    silencing of their voices, 31
    socialization, 4
Goldberger, N. R., 19
Goldstein, L. S., 86
Graziano, A. M., 16
Green, T. F., 66
grounded theory, 68
Grumet, M. R., 30
guided participation, 5

Harding, S., 17
Harter, S., 127
Haug, F., 2, 4, 25, 27, 31, 34, 36, 37, 38, 67,
    71, 133, 139
    recursive nature of memory-work, 35
    theory of memory-work, 5
    theory of socialization, 25
Hayward, J., 91, 139
Helgesen, S., 120
Hermanowicz, J. C., 84
Hewitt, N. M., 21, 22
Hill, S. T., 21
Hoagland, S. L., 127
Hofstra University, 27
Holden, C., 15
Holton, G., 23

Hubbard, R., 20, 47, 65

imagination, 4, 90–91
individual science, 131
"inside science," 106
intellectual tools of childhood, 98
interruptions, 8, 132
Isle of Skye (Scotland), 91

Jagger, A. M., 63
Jarvis, T., 14
Jayaratne, T. E., 18
Johnson, M., 45, 66, 78, 79, 133

Kanter, R. M., 120
Keller, E. F., 6, 17, 18, 140
Kermode, F., 137
Kippax, S., 2, 30
Koch, J., 138
Kotre, J., 37
Kreisberg, S., 116–17, 120, 125, 130
Kuhn, T. S., 17, 82

Lahar, S., 98, 138
Lakoff, G., 66, 78, 79, 133
Legerstee, M., 46
Leslie, L. L., 22
Lewis, R., 95, 98
literary criticism, 66
lived distance, 5, 64, 86
Lutz, D., 14, 24
Lyotard, J. F., 19

Macilwain, C., 15
Marxism, 4
Maynard, M., 18, 63
Mayr, E., 15
McClintock, B., 17, 140
McClure, G. T., 22
memory
    accuracy of, 41
    amalgam, 34, 35
    analyzing sets of, 38
    and "clever disobedience," 107–12, 135
    clichéd, 30
    as a construct, 39–42
    developmental perspective of, 37
    episodic, 34

event, 34
and face-saving, 111
and the integration of the senses, 48–51
    hearing, 54–56
    smell and taste, 60–61
    touch, 56–60
    vision, 52–54
and lack of control, 118
and metaphors, 67, 69
and missing elements, 37
and play, 102
power in, 117–19
of nature, 99
and reflection, 109
semantic, 34
the sensuous in, 46, 64
traditional research on
    Also see memory-work
memory-work
    and autobiography, 27, 29
    and biography, 27, 29
    and censored memories, 32
    collective analysis, 30, 37
    and creativity, 7
    and creativity in play, 85
    definition of, 2
    decision about which memories to study,
        30–31
        refraining from judgment, 33
    dominant control issues, 120
    and earliest childhood memories, 2
    emergent analysis, 38–39
    and families, 8
    and family differences, 112–13
    feminist-based, 25
        and embodiment, 63
    format of memories, 31
    influences on research, 2
    and limits of language, 30
    and Marxism, 4
    methodology of, 7, 27–28
    and narrative inquiry, 28–30
    and nature, 132
    and "personal science," 3
    and play, 87–90
    and power, 8
    recursive nature of, 35
    rewriting memories, 31

memory-work (*continued*)
  rules, 31
  and sensuous experience, 7
  and social construction of self, 27
  and socialization in relationship to nature
      and science, 2, 6
  strategies of memory generation, 32
  subject and object inseparability, 5
  theoretical context of, 4–6, 7
  therapy, 27, 102
  traditional analysis, 35–37
  types of memories, 34
  Also see memory
Merchant, C., 18, 138
Merleau-Ponty, J., 5
metaphor, 7, 66, 133, 134
  and elements, 70–71, 72–73, 75–76, 133
  experience through, 66
  making meaning from, 70
      efficacy, 72–74, 133
      fear, 70–72, 133
      growth and development, 78–80, 133
      relationship, 75–77, 133
  and memories, 67, 69
  rituals, 79–80
  role in science, 82
  as a safe outlet for divergent views, 70
  tree, 76–77, 79
  uses of, 67–70
  variation from culture to culture, 67
Milk, H., 114
Miller, J. B., 86, 116, 120, 125
Mills, T. J., 135
Mistry, J., 5
Morss, J. R., 88, 138
Moscone, G., 114
Mosier, C., 5
Muir, J., 138
multiple sensory avenues, 49

Nabhan, G. P., 89
Nachmanovitch, S., 88
narrative justification, 136
narrative and memories, 53
National Science Foundation, 16, 21, 22
nature
  and culture, 98
  distance from, 64, 80–82, 113, 132

engagement with, 59
and family, 99, 113
fathers' influences on memories of nature,
    99–105
mothers' influences on memories of nature,
    105–107
and personal science, 138–39
Also see memory-work
Nesselroade, J. R., 15
Neumann, W. L., 15
new meaning
  and creativity, 85
Nobel Prize, 17
Noddings, N., 28
Norman, D., 39

Oaxaca, R. L., 22
obedience quotient (OQ), 21
Oklahoma State University, 27
Onyx, J., 2
organic cues, 31
outlaw emotions, 63
Overton, W. F., 4, 16, 17
Oxford English Dictionary, 125

Pathways in Science Project, 22
personal science, 3, 47–48, 133
  active observation, 51
  development of, 61–65
  in the domestic sphere, 48
  exploration and experimentation, 92–94
  and nature, 138–39
  power in, 130
  role of metaphor in, 82
  use of, 135
Petrie, H. G., 66
phenomenology, 64, 138
Piirto, J., 89
play, 87–90
  abandonment of, 88
  and adult creativity, 89
  and experimentation, 6
  and imagination, 90–91
  and memory, 102
  rhetorics of, 88
play spaces, 24
Porter, T., 70
positivism, 6, 16, 17

five assumptions of, 18
power, 8, 134
  and apprenticeship, 117, 123–25, 130
  as competence or mastery, 125–29
  and control, 118–19
  definition of, 116
  as domination, 129
  etymology of, 125
  exchanges of, 122–25
    apprenticeships within families, 123–25
    taking turns in career decisions, 122
  feminists writing on, 116
  of girls, 116
  in memories, 117–19
  in personal science, 130
  and women, 117
    in relationships, 120–22
Prometheus complex, 73
Psychoanalysis of Fire, The (Bachelard), 73

Raulin, M. L., 16
Rayman, P. M., 21
recursive nature of memory-work, 35
Reese, H. W., 15
reflection, 109
resistance, 112
Rich, A., 9, 41, 42
Rogoff, B., 4, 5, 117, 123, 125, 131
Rose, S., 15
Rosener, J. B., 120

Sacks, O., 52
Sadker, D., 31
Sadker, M., 31
Schmuck, P., 120
Schram, P. J., 17
Schratz, M., 27, 28, 68, 102, 105
Schumaker, P., 116, 120
science
  character of, 13
  children's first impression of, 13, 14–15
  cultural grounding in, 47
  definition of, 16
  and efficacy, 80
  elite, 6, 16
  feminist critiques of, 14, 17–19
  and relational thinking, 17
  role of metaphor in, 82
  in school, 80, 82–84
  making sense, 45
  and social bias, 17
  views of, 15–17
  women in, 7
  women's views of, 19–20
scientific concept, Vygotsky, 4, 24
Scott, A., 63
secrecy, 137
sediment, metaphor of, 34
self-efficacy, 22
senses, 45
  exploration of, 58–59
  hearing, 54–56
  integration, 48–51
  perception and social interaction, 46
  smell and taste, 60–61
  touch, 56–59
  vision, 52–54
sensuous, the, 6, 7
  connections, 45
  and embodiment, 38, 45, 56
  importance of, 133, 135
Seymour, E., 21, 22
Shiffrin, R. M., 39
small spaces, 89, 90, 91, 134
Smolucha, F., 5, 89
Snark Syndrom, 20
social bias, 17
social constructivism, 4
social interaction, 46
social signaling, 54
Sonnert, G., 22
Spanier, B., 19
spontaneous concept, Vygotsky, 4, 24, 64
Starhawk, 129
stereotypical scientist, 14
Stewart, A. J., 18
Stiver, I. P., 86
Strahler, A. N., 3
Sullivan, A. M., 31
Sutton-Smith, B., 87, 88

Tarule, J. M., 19
Taylor, J. M., 31
theory of absorption, 106
theory of ecofeminism, 138
theory of memory-work, 5

theory of socialization, 25
therapy and memory-work, 102
Thoreau, H. D., 138
Tobias, S., 17, 22
"Transcendental Etude" (Rich), 8–9
Trevarthen, C., 46
Trimble, S., 89
Tulving, E., 34, 39
Turner, M., 66

U. S. National Science Foundation, 16

van Oech, R., 89
von Hornbostel, E. M., 61
Vygotsky, L. S., 4, 5, 24, 59, 64, 88, 89, 92, 95

Walker, R., 27, 28, 68, 102, 105
Watson, J., 15
Wellesley College, 22
Welty, E., 29
Wertsch, J. V., 4

Winnicott, D. W., 85, 87
Winterson, J., 138
Witherell, C., 28
women
   difficulties of combining careers and parenthood, 22–23
   enrolled in engineering and science programs, 20
      dropout statistics, 21
   and relational thinking, 17
   and power, 117
      in relationships, 120–22
   relationship to nature, 13
   role ambiguity, 23
   underrepresentation in science, 20
   views of science, 19–20
Women's Ways of Knowing (Belenky et al.), 19
Wynne, B. E., 20

zone of proximal development, 5

Studies in the Postmodern Theory of Education

General Editors
Joe L. Kincheloe & Shirley R. Steinberg

Counterpoints publishes the most compelling and imaginative books being written in education today. Grounded on the theoretical advances in criticalism, feminism, and postmodernism in the last two decades of the twentieth century, Counterpoints engages the meaning of these innovations in various forms of educational expression. Committed to the proposition that theoretical literature should be accessible to a variety of audiences, the series insists that its authors avoid esoteric and jargonistic languages that transform educational scholarship into an elite discourse for the initiated. Scholarly work matters only to the degree it affects consciousness and practice at multiple sites. Counterpoints' editorial policy is based on these principles and the ability of scholars to break new ground, to open new conversations, to go where educators have never gone before.

For additional information about this series or for the submission of manuscripts, please contact:

Joe L. Kincheloe & Shirley R. Steinberg
c/o Peter Lang Publishing, Inc.
275 Seventh Avenue, 28th floor
New York, New York 10001

To order other books in this series, please contact our Customer Service Department:

(800) 770-LANG (within the U.S.)
(212) 647-7706 (outside the U.S.)
(212) 647-7707 FAX

Or browse online by series:
www.peterlangusa.com